WORLD LEISURE PARTICIPATION
Free Time in the Global Village

WORLD LEISURE PARTICIPATION
Free Time in the Global Village

A project of the World Leisure and Recreation
Association Commission on Research

Edited by

Grant Cushman

Lincoln University, New Zealand

A.J. Veal

University of Technology, Sydney, Australia

and

Jiri Zuzanek

University of Waterloo, Canada

CAB INTERNATIONAL

CAB INTERNATIONAL   Tel: +44 (0)1491 832111
Wallingford   Fax: +44 (0)1491 833508
Oxon OX10 8DE   E-mail: cabi@cabi.org
UK   Telex: 847964 (COMAGG G)

A catalogue record for this book is available from the British
Library.

ISBN 0 85198 975 6

Typeset by York House Typographic Ltd, London
Printed in the UK at the University Press, Cambridge

Contents

Preface

The idea for this book was initially mooted at the first conference of the Research Commission of the World Leisure and Recreation Association (WLRA), held in Twannberg, Switzerland, in 1982. Major leisure participation surveys have been conducted in many countries around the world since the 1960s, involving hundreds of thousands of respondents and substantial expenditure. The idea was formulated that the experiences and results arising from this major research effort should be examined, compared and evaluated. It is likely that such surveys will continue to be conducted in the economically advanced countries, and that developing countries will consider conducting surveys as leisure becomes increasingly significant as a result of economic development. It would therefore be valuable for all concerned if the accumulated wisdom arising from the surveys conducted to date could be consolidated, so that methodology could continue to be improved and past mistakes avoided in future.

The idea of undertaking comparative analysis of national leisure participation surveys was discussed at subsequent congresses in the United Kingdom, France, Australia and Germany, organized under the auspices of WLRA, the International Sociological Association (Research Committee 13) and the UK Leisure Studies Association. While no funding was forthcoming to support the project, CAB INTERNATIONAL generously offered to publish a book addressing this topic.

The first criterion for including a contribution in this book was that the country involved had to be, broadly speaking, economically developed. In practice few developing countries have the resources to conduct leisure participation surveys; it is largely an activity of industrial, economically developed, relatively wealthy countries. There are exceptions, and there is increasing interest in leisure and tourism issues among developing countries. Future editions of the book will therefore be more inclusive.

The second criterion for inclusion was that at least one national leisure participation survey had been conducted. Third, it was necessary to identify a researcher who was willing and able to write a summary of that country's experience with leisure surveys, in English.

Invitations to contribute were published in various newsletters which circulate in the academic community and a number were directly solicited through personal contacts. A number of potential contributions, in addition to the twelve included, were promised but unfortunately failed to materialize. Again, it is to be hoped that future editions may remedy this.

The prospect of being able to compare survey results from different countries is a distant one; differences in survey methodology and in definitions and attitudes arising from cultural diversity make the results of existing surveys virtually impossible to compare. Therefore, it is not possible to state with any confidence that, for example, one country is more active than another in sport or in the arts. It is hoped that this book will assist in raising awareness among those responsible for national surveys that there are others in the world engaged in the same activity so that, in time, comparable features will begin to be built into survey designs.

In future editions of this book we hope to update the survey data from the twelve countries represented and to extend the coverage to more countries. Furthermore, we would hope to be able to report on increasing levels of comparability between national surveys.

We would be interested to receive offers of contributions for future editions of the book.

G.C.
A.J.V.
J.Z.
October 1995

About the Contributors

James J. Bason is Research Coordinator in the Survey Research Center, University of Georgia, Atlanta, Georgia, USA.

H. Ken Cordell is Project Leader and Supervisory Research Forester for Outdoor Recreation and Wilderness Research, USDA Forest Service, USA.

Grant Cushman is Professor of Parks, Recreation and Tourism and Director of the Resource Studies Division at Lincoln University, New Zealand.

Simon Darcy is lecturer in the School of Leisure and Tourism Studies, University of Technology, Sydney, Australia.

Chris Gratton was, at the time his contribution was written, Professor of Leisure Studies at Tilburg University, The Netherlands, but has since taken up a post in the School of Leisure and Food Management, Sheffield Hallam University, U.K.

Munehiko Harada is an associate professor in sport and recreation management at the Osaka University of Health and Sport Sciences in Japan.

Bohdan Jung is Professor, Research Institute for Developing Economies, Warsaw School of Economics, Warsaw, Poland.

Allan Laidler is Senior Lecturer and Course Coordinator in the Recreation and Leisure Studies programme at the Victoria University, Wellington, New Zealand.

Burt Lewis is Research Technician, University of Georgia, Atlanta, Georgia, USA.

Barbara L. McDonald is Social Scientist, Forest Inventory, Economics and Recreation Research, USDA Forest Service, Athens, Georgia, USA.

Conceptión Maiztegui-Oñate is in the Faculty of Philosophy and Educational Sciences, University of Deusto, Bilbao, Spain.

Jack K. Martin is Director, Survey Research Center, University of Georgia, Atlanta, Georgia, USA.

Harald Michels is a Sport Scientist and Lecturer in the Department of Leisure Science of the German Sport University in Cologne, Germany.

Morgan P. Miles is Associate Professor of Marketing, Department of Marketing, Georgia Southern University, Statesboro, Atlanta, USA.

Hillel Ruskin is Professor of Physical Education and Chairman of the Cossel Center of Physical Education, Leisure and Health Promotion and Director of the Graduate Program of Physical Education in Public Health of the Faculty of Medicine, Hebrew University, Jerusalem, Israel.

Bob Robertson is Dean of the Faculty of Business and Professor of Leisure and Tourism Studies at the University of Technology, Sydney, Australia.

Nicole Samuel is Head of the CNRS research team: Temps Sociaux, Ages et Modèles Culturels, Paris, France.

Atara Sivan is Assistant Professor in the Department of Education Studies at Hong Kong Baptist University.

Walter Tokarski is Professor of Sport and Head of the Department of Leisure Science of the German Sport University in Cologne, Germany.

A.J. (Tony) Veal is Associate Professor in the School of Leisure and Tourism Studies, and Director of the Centre for Leisure and Tourism Studies at the University of Technology, Sydney, Australia.

Jiri Zuzanek is Professor of Leisure Studies in the Department of Leisure Studies, University of Waterloo, Ontario, Canada.

List of Tables and Figures

TABLES

FIGURES

1 National Leisure Participation Surveys: An Overview

GRANT CUSHMAN, A.J. VEAL AND JIRI ZUZANEK

INTRODUCTION

How many people play sport in the world today? What are the trends in television watching? Who patronizes the arts? Who visits national parks? Despite the increasing global significance of sport, entertainment, culture and the conservation and use of the environment in cultural and economic life, it is not possible, at present, to answer these simple questions on worldwide patterns of leisure participation. While international data are collected on such phenomena as health, housing, education and economic activity,[1] no comparable data exist for leisure activity. It might be thought that leisure is not sufficiently important to justify the cost of gathering such information. In fact, leisure is one of the human rights safeguarded by the United Nations Universal Declaration of Human Rights, which, in Article 24, states:[2]

> Everyone has the right to rest and leisure, including reasonable limitation of working hours and periodic holidays with pay.

Rights in relation to the arts and culture are recognized in Article 27:

> Everyone has the right freely to participate in the cultural life of the community, to enjoy the arts and to share in scientific advancement and its benefits.

In relation to one aspect of leisure, the Council of Europe, in its *Sport for All Charter*, states:

> Every individual shall have the right to participate in sport.

In 1987, the then Secretary-General of the United Nations declared:

> One of the primary needs of the human person is leisure and such use of it as will provide psychological strength and refreshment.
>
> (Perez de Cuellar, 1987)

Thus leisure has been recognized by national governments and international organizations as being of sufficient importance to be accorded the status of a human right and a human need. This is tacitly recognized by the involvement of governments at national, regional and local levels throughout the world, in such spheres as sport, physical recreation and education, outdoor recreation in urban and natural areas, the arts, heritage, broadcasting and children's play. It is widely accepted that leisure activity, in various forms, makes a significant contribution to the quality of life of individuals and communities (Marans and Mohai, 1991). On the other hand there are aspects of leisure that can be harmful, such as abuse of legal and illegal drugs, sporting accidents, and activities that cause environmental or cultural degradation.

This book has been designed to draw together information on leisure participation in a number of developed countries. The data from the various countries represented were collected at different times, using widely differing methodologies, so possibilities for comparison are very limited. However, they all reflect a worldwide growing recognition of the importance of leisure to societies, economies and environments.

Like so many aspects of modern life, leisure is being affected by global forces. This can be seen in trends in groups of activities as diverse as home-based leisure, sport, entertainment and tourism. Most leisure takes place in the home, but increasingly the leisure 'products' which people consume in their homes are being produced and distributed on a worldwide scale, including film and television programmes, recorded music, books and magazines and computer games. Sport, which was once primarily a neighbourhood leisure activity, is now a global one. This is reflected not only in such overtly international phenomena as the Olympic Games and other international sporting championships, but in the internationalization of sporting 'culture' via the media and the marketing of clothing and equipment. Modern entertainment industries have always been international in nature, including film, popular music and television. The ultimate international leisure activity is tourism, now seen as one of the world's largest and fastest growing industries. Linked to tourism are the phenomena of the natural environment and cultural heritage, which are the basis of much local and international tourism, giving rise to issues of 'ownership' and conservation.

Appadurai (1990) suggests that we can conceive of five dimensions of global cultural flows. First, there are *ethnoscapes* produced by flows of people: tourists, immigrants, refugees, exiles and guestworkers. Second, there are *technoscapes*, the machinery and plant flows produced by multi-national and national corporations and government agencies. Third, there are *finanscapes*, produced by the rapid flows of money in the currency markets and stock exchanges. Fourth, there are *mediascapes*, the repertoires of images and information, the flows that are produced and distributed by newspapers, magazines, television and film. Fifth, there are *ideoscapes*, linked to flows of images that are associated with state or counter-state movement ideologies which comprise elements of the Western Enlightenment world-view images of democracy, freedom, welfare, rights, etc.

This globalization process, with its associated processes of exchange, circulation and commodification which characterize modern market societies (Rojek, 1995, p. 92; Jarvie and Maguire, 1994, p. 230), points to increasing global interrelatedness. This can be understood as leading to global ecumene, defined as a 'region of persistent culture interaction and exchange' (Kopytoff, 1987, p. 10; Featherstone, 1990) and is witnessed by a growing internationalization of ideas and consumer expectations in leisure and recreation (Mercer, 1994).

Leisure is therefore increasingly an intrinsically international phenomenon, but there are also trends and challenges in leisure at the local level which recur around the world and where lessons can be learned from the sharing of experience. Thus the role of leisure activity in the lives of young people and the elderly is an issue being addressed in many countries. Governments around the world are involved in promoting particular forms of leisure activity, for example sport and the arts. Questions arise as to what organizational and promotional strategies are most effective in these areas. The redevelopment of dockland areas for leisure purposes and the link between leisure provision and city 'boosterism' are again issues that arise throughout the developed world. Environmental and cultural conservation issues, arising from local recreational pressures as well as international tourism, are also problems faced by many societies. In many countries questions are being raised about how leisure amenities which have traditionally been provided by the public sector, such as parks and swimming pools, should be managed: by the public sector, the voluntary sector, the commercial sector, or various combinations. These, and many other examples, suggest that leisure throws up problems and issues that are far from unique. Some sharing of information and experience therefore seems desirable.

This book concentrates on economically developed countries. In general, third world or developing countries have not had the resources to devote to substantial research on leisure. Leisure is not, however, an exclusive concern of the developed world; it is present in various forms in all cultures and at all stages of economic development. Most music and drama has deep cultural roots, modern sports were preceded by centuries-old local and regional sporting contests, and massive religious pilgrimages long pre-dated modern mass tourism. Such issues as the need for open space for recreation in city environments, the role of sport in health and national prestige, the status of indigenous culture in the face of mass media influences, and the problems of conservation of natural and historic heritage in the face of growing population and tourist pressures, are, if anything, more pressing in developing countries than in the developed world. Joffre Dumazedier (1982) suggested that the 'comparative method' might be used as a method for examining possible social futures – that is, one society can examine alternative futures for itself by studying the experiences of others, particularly those which are more economically developed. This can apply among economically advanced countries, since there is enormous variation in wealth and social practices even among members of the developed 'club', but it is particularly appropriate between developed and developing countries. The experiences of the handful of economically developed countries presented in this volume, especially when read in conjunction with some of the more policy-orientated volumes reviewed below, will therefore, it is hoped, be of interest to others as they consider issues of leisure policy and leisure development within their own countries.

INTERNATIONAL DATA ON LEISURE

While this book is the first attempt to bring together cross-national data on leisure participation generally, it reflects a growing interest in cross-national research (see for example, Hantrais and Samuel, 1991 and Aitchison, 1993), and has a number of predecessors in the form of data collections related to particular aspects of leisure, and it relates to a substantial tradition of general international comparative research. In this opening chapter therefore, before outlining the scope and nature of the book, these previous data collection exercises are briefly reviewed and the problems of conducting international, or cross-national, comparative research are considered.

Data are regularly collected on an international basis for various aspects of leisure which are seen as *industries* and where data are routinely gathered by national governments for taxation and other legalistic purposes. For example, the World Tourism Organization and the Organization for Economic Cooperation and Development both assemble data on international travel, most of which is for leisure purposes (WTO, Annual; OECD, Annual). The Universal Declaration of Human Rights indicates the concern of national governments with working hours, which can be seen as the obverse of leisure time; the UN's International Labour Office gathers data on working hours and paid holidays in member countries (ILO, Annual). And data exist on levels of ownership of some leisure goods, for example television sets. Because of universal national licensing and taxation policies, data are readily available on alcohol and tobacco consumption.

However, for more comprehensive information on leisure participation it is necessary to conduct special social surveys of the population. Surveys can take two forms: activity-based and time-based. Activity-based surveys include the national leisure participation surveys which form the basis of this book; these surveys use questionnaires to gather information on participation in specified leisure activities over a specified period of time. Time-based surveys are sometimes called 'time-budget' or 'diary' studies; they involve respondents keeping *diaries* of their activities over a specified period of time, perhaps one or two days. Start and finish times of activities are recorded, including simultaneous activities (for example, listening to the radio while eating) and sometimes location and company. Demographic and socioeconomic data are usually gathered also. The responses are then coded for analysis purposes. The result is generally a very substantial amount of data, given the number and range of activities that people can engage in over a 24-hour period. While activity-based surveys are probably the most common, in some countries, time-based surveys have taken precedence.

In the field of leisure surveys there has been only one example of truly multinational cooperation to produce data that could be compared between countries, and this was a time-based study. The *Multinational Comparative Time-Budget Research Project* was conducted in 13 countries (see Table 1.1) in the mid-1960s (Szalai, 1972; Feldheim and Javeau, 1977). The study was coordinated through the UNESCO-funded European Coordination Centre for Research and Documentation in Social Sciences in Prague, but was conducted by a variety of academic and governmental organizations in the participating countries. Leisure is of course only one use of time, so this study was only partially concerned with leisure. The study established the considerable similarity in the daily patterns of life of the populations of the 13 countries studied, the major differences arising from variations in the economic status of women and the differences in the extent of television ownership which existed at the time (Feldheim and Javeau, 1977). Clearly both of these factors will have changed substantially over the intervening 25 years, but no follow-up study has been attempted, although time-budget studies continue to be conducted in many countries.

Under the auspices of the International Sociological Association, three volumes of collected papers on aspects of leisure in a number of countries have been produced, covering the arts, sport and 'lifestyles'.

Trends in the Arts: A Multinational Perspective (Hantrais and Kamphorst, 1987) included contributions from eight countries (see Table 1.1). Published in 1987, the book is based largely on data collected in the 1970s and early 1980s. The extent and quality of data available vary significantly from country to country. In addition to information on participation levels, data on attendances at arts venues and funding of arts organizations are also presented. One of the main conclusions drawn in the book is that participation in the arts is concentrated among the more highly educated, higher income groups in society in most of the countries studied, and the editors call for more

Grant Cushman et al.

research, including qualitative research, to explore the processes that lead to this widespread situation.

Trends in Sports: A Multinational Perspective (Kamphorst and Roberts, 1989) includes contributions from 15 countries (see Table 1.1). Despite the enormous variety of social, physical and economic environments represented by the 15 countries, the editors conclude that the universal nature of sporting

Table 1.1. Countries involved in five cross-national leisure projects.

	Time-budget study, 1972*	Trends in the arts, 1987†	Trends in sport, 1989‡	Life-styles, 1989§	Leisure policies in Europe 1993¶	Current book
Australia						✓
Belgium	✓			✓		
Brazil				✓		
Bulgaria	✓		✓			
Canada		✓	✓			✓
Czechoslovakia	✓	✓	✓			
Finland			✓			
France	✓	✓	✓	✓	✓	✓
Germany (GDR)	✓					
Germany (FDR)	✓				✓	✓
Great Britain		✓	✓	✓	✓	✓
Greece					✓	
Hong Kong						✓
Hungary	✓	✓		✓		
India			✓			
Israel						✓
Italy			✓			
Japan			✓	✓		✓
Netherlands	✓	✓	✓		✓	
New Zealand			✓			✓
Nigeria			✓			
Peru	✓					
Poland	✓		✓	✓	✓	✓
Portugal			✓			
Puerto Rico		✓		✓		
Spain					✓	✓
Sweden					✓	
USA	✓	✓	✓	✓		✓
USSR	✓				✓	
Yugoslavia	✓					

* Szalai, 1972; Feldheim and Javeau, 1977; † Hantrais and Kamphorst, 1987; ‡ Kamphorst and Roberts, 1989; § Olszewska and Roberts, 1989; ¶ Bramham *et al.*, 1993.

participation is clear; that walking, running and swimming make up the bulk of sporting activity across the world; that there are widespread common perceptions of the importance of sport in modern social life; and that it is even possible to detect trends in participation, particularly the rapid growth of participation in the 1960s and 1970s and the slowing of growth in the 1980s.

Leisure and Lifestyle: A Comparative Analysis of Free Time (Olszewska and Roberts, 1989) contains contributions on nine countries (see Table 1.1). While some of the contributions present survey data on leisure participation, overall they are less concerned with presentation of data than with painting a sociopolitical picture of the context of leisure in each country, particularly in the context of the economic recession of the late 1980s.

Similarly, *Leisure Policies in Europe* (Bramham *et al.*, 1993), as the title implies, is concerned with policies rather than data on participation. The contributions, from nine countries (see Table 1.1), are set against the background of the 'new world order', involving the transformation of Eastern Europe, including the collapse of the Soviet Union, the expansion of the European Community and the emergence of an international, post-industrial, post-Fordist, post-modern European society.

Participation by various countries in the above projects and in the current book is partly fortuitous, reflecting networks between individual researchers, chance meetings at conferences, synchronization of projects and personal and organizational time and resources, and the availability of facilities to overcome language barriers (all the projects have used English as the common language). But it also reflects, to some extent, the degree to which the various participant countries are involved in leisure policy and research, particularly survey-based research. It therefore seems worthwhile to summarize the pattern of country participation, as shown in Table 1.1. Altogether 30 countries have been involved in these projects, but it is notable that only one, France, has been involved in them all, although the USA took part in all except the European project. Of the 30 countries, 14 have been involved in only one of the projects. It is notable that there has been no involvement from Africa, from mainland South America or from South-east Asia. While Europe is heavily represented, the absence of some of the most affluent countries, especially Switzerland, Austria, Norway and Denmark, is notable.

Only the first of the studies reviewed above was designed from the beginning as a cross-national project; the rest have relied on the use of existing data sources. While they reflect the diversity of experience in the 30 countries involved they also reflect a common theme: a growing interest in leisure, both as a social phenomenon and as an issue for public policy. It is notable that the later volumes among those reviewed include the word *leisure* in their titles. There would appear to be a growing consensus that *leisure* is a valid focus for international comparative study.

Again with the exception of the *Comparative Time Budget* study, all these projects illustrate the *lack* of comparability between nation states in terms of data sources, research traditions and administrative arrangements. One of the long-term aims of this book is to stimulate discussion on how countries conducting leisure participation surveys might move towards more comparability in future surveys so that comparisons and aggregation might be achieved. Leisure surveys are, however, not alone in facing problems of comparability: cross-national comparative research generally is faced with a myriad of problems.

CROSS-NATIONAL COMPARATIVE RESEARCH

Why conduct cross-national comparative research? One goal might be to produce 'league tables' – to show where different countries stand in relation to one another, to address what Novak (1977) has termed *nation-oriented problems*. This is done continually in relation to such phenomena as economic growth rates, per capita income levels, taxation levels and crime rates, and these 'league tables' are often reported in the media. But implicit in even these comparisons, are quasi-scientific questions and answers – in Novak's terms, *variable-orientated problems*. Media and political analysts do not usually refer to such data without also providing some commentary; they usually seek 'explanations', often in order to pursue a particular political line of argument. For example rates of economic growth are often causally linked to levels of taxation or levels of expenditure on education or research in different countries, or crime rates are linked to levels of gun ownership or inequalities in wealth distribution. Implicit in such statements are causal models, relating one variable to another. The comparisons between nations become part of the political and social discourse. As Warwick and Osherson put it: 'Rather than being a second-order activity tacked onto more basic cognitive processes, comparison is central to the very acts of knowing and perceiving' (1973, p. 7).

Warwick and Osherson (1973, pp. 8–11) explain how cross-national comparative research can contribute to theory building and theory testing in social research. Cross-national comparative research, they argue, helps in developing 'clearly defined and culturally salient concepts and variables'; it enables theories to be tested against a wider range of conditions than single-nation research, enabling a greater degree of generality to be achieved; and it is good for the researcher, developing a 'heightened sensitivity to the differential salience and researchability of concepts in varying cultural settings'. In leisure research it is therefore salutary to be reminded that not all countries, even in the economically developed world, experience recession at the same time or in the same way. Some countries are still developing industrially, while others are experiencing de-industrialization; some are grappling with the problem of a growing youth population, while others are experiencing an ageing of the population. While some theories about leisure may be intentionally confined to particular social, economic, geographical and political milieux, there must surely be some desire among the research community to develop theory which has a wider application.

The benefits of cross-national comparative research are easy to identify: actually achieving the benefits is more difficult. Numerous difficulties are presented to researchers attempting to overcome cultural and language barriers to conduct cross-national research. While researchers may struggle to overcome them, it is in such differences that the value of cross-national comparison may lie. As Przeworski and Teune put it:

> To say that a relationship does not hold because of systematic or cultural factors is tantamount to saying that a set of variables, not yet discovered, is related to the variables that have been examined.
>
> (1973, p. 123)

Nevertheless there must be variables in common across countries for cross-national research to be of value. We have barely begun to address these issues in cross-national leisure research. For example, surveys may indicate certain levels of participation in 'football' in various countries, but 'football' includes a variety of different sports, including American grid-iron, soccer, rugby league and rugby union, and Australian Rules and Gaelic football. Some of these codes are 'national' sports, some are regional and some are very much minority sports: thus the term 'football' implies a wide variety of phenomena, rather than a single one. In many countries much leisure activity revolves around the consumption of alcohol, while in others it is forbidden. In some countries gambling is part of the culture, while in others it is frowned upon or banned. Such fundamental differences might eventually lead to the identification of a range of 'functional' activities for comparison in cross-national studies – for example the idea of a 'national sport' or 'focal cultural activity'.

But this assumes that comparisons can be undertaken at all. Much of the literature on cross-national comparative research focuses on the *design* of data collection projects designed from the outset to be conducted cross-nationally. In the leisure area, apart from the *Multinational Comparative Time-Budget* study, researchers have had to be content with making comparisons using data already collected in separate countries, at different times and for different purposes. Any level of comparison at all is therefore problematical. This is further confounded by the growing realization, resulting from experience with single-nation surveys, that leisure participation data are extremely sensitive to the methodology used in their collection. Responses are affected by the differing definitions of *leisure* itself, the age-range of respondents included in surveys, the time of year when data are collected, and the participation 'reference period' used. These issues are considered in more detail in the concluding chapter of the book.

THE ORIGINS AND ROLES OF LEISURE PARTICIPATION SURVEYS

The modern era of leisure research began in the 1960s, particularly with the work of the United States Outdoor Recreation Resources Review Commission (ORRRC, 1962), which involved, among other techniques, the conduct of large-scale national surveys to establish base data on levels of participation. Other western countries rapidly followed suit with their own surveys. The impetus for these studies of leisure activity was a combination of growing affluence, the increase in car ownership and consequent growth in car-based recreation, and the growth of the population, particularly the young 'baby boomer' population, in most western countries. At that time large-scale, national or regional, questionnaire-based community leisure participation surveys vied with on-site surveys of users of individual recreation facilities or networks of facilities (usually outdoor recreation areas) as the main vehicle for empirical data collection. The initiative and resources for the studies came largely from governments and government agencies, driven by policy concerns, while academics and consultants were involved as advisers and as primary and secondary analysts.

These early surveys were generally purely descriptive – that was their purpose. Whether governments of the day were concerned about outdoor recreation, sport and physical recreation or patronage of the arts, they needed data in order to formulate, refine or monitor policies. In some cases the data were required to provide an initial 'position statement' – for example on the proportion of the population engaging in sporting activities and the variation in participation levels among various social groups. In other cases, in situations of rapid economic and demographic growth, the data provided the basis for demand forecasting. These early surveys can be seen as part of a general concern for social policy issues which was a feature of interventionist western governments of the 1960s. In the former communist bloc they reflected government aims to establish 'socialist lifestyles' and to research ways and means of doing this (Filipcova, 1972). While western governments have generally become less interventionist over the intervening 30 years, many of the institutions established in the 1960s to administer government policies on various aspects of leisure, have continued to generate a demand for data on leisure participation.

In the area of social behaviour the 'facts' are continually changing. In contrast, in most of the physical sciences a discovery, once made, is forever true – even though its theoretical explanation may change. In the social sciences a discovery may be true only for the instant in which it is made; from that time on its value as a description of contemporary society begins to 'decay'. It therefore becomes necessary to update such data continually. This is certainly true of data on patterns of leisure participation. Indeed, it is the actual and potential fluidity of leisure behaviour that often gives rise to the need for data collection in the first place. Changes in leisure participation arise from social, economic and environmental influences, such as changes in personal incomes, in social values or in technology. Governments, and other organizations, seek to anticipate and monitor the changes, particularly when they demand a policy response. For example, government agencies have to cope with any increased demand for recreation on public lands. In other situations governments and other agencies seek to stimulate change them-selves – for example in promoting sports participation and exercise or partici-pation in the arts. In these cases data are required to monitor trends and to assess the effects of policy measures.

So, while they have had a chequered career, where governments have an interest in leisure phenomena and where the resources are available, periodic surveys of leisure participation have become the norm.

SURVEYS IN THE LEISURE STUDIES CONTEXT

In addition to their policy roles, early leisure participation surveys laid the groundwork for the development of a variety of traditions in leisure studies. Researchers in the USA developed traditions based on quantitative modelling and demand prediction and on quantitative behavioural models at the individual/psychological level (Cichetti, 1972). These traditions have tended to be prominent in leisure research in North America ever since. In the UK and Europe the quantitative/modelling approach was soon abandoned in favour of a more direct use of such data in policy formation and monitoring.

Theoretical and critical researchers in the leisure area, while generally eschewing the survey method, have nevertheless often drawn on the evidence of survey data as a starting point for their analyses, particularly in relation to social class and gender differences in participation levels, and in relation to publicly subsidized areas of leisure such as élite sport, the arts and outdoor recreation. As leisure studies have grown as an area of tertiary study, leisure participation survey data have found a role in textbooks and in the classroom, in providing students with an empirical picture of leisure participation pat-terns. Theoretical and critical research on leisure tends to be conducted by academics rather than government agencies. But academics in the social sciences – and in leisure studies in particular – do not themselves generally have access to resources to conduct large-scale empirical research and so have been reliant on government sponsored surveys when discussing general patterns of leisure behaviour.

Recently surveys have had a 'bad press' from academics, particularly in light of the growing popularity – and indeed orthodoxy – of qualitative research methods in the field. In order to establish the case for undertaking or placing more emphasis on qualitative empirical research and non-empirical theoretical research, commentators often outline the shortcomings of quantitative methods, including surveys (Clarke and Critcher, 1985, pp. 26–27; Rojek, 1989, p. 70; Henderson, 1991, p. 26; Aitchison, 1993). The critics often impute motives and attitudes to researchers who utilize survey methods, implying that they are somehow wedded to a somewhat outdated and extreme version of 'positivism' to the exclusion of other research approaches. The cumulative effect of repeated detailing of their failings and limitations has been to put surveys in a bad light with some of the leisure research community, and to create a 'phoney war' between alternative methodologies. The survey method has strengths and limitations, as do all research methods. For example, for the survey method, making definitive *descriptive* statements about the community as a whole is routine, but *explanation* of observed behaviour is often speculative at best. Conversely qualitative methods are often strong on explanation but relatively weak with regard to reliable generalization to the wider community. Thus survey methods, other forms of quantitative method and qualitative research methods should, in our view, be seen as complementary (Kamphorst *et al.*, 1984).

Some of the implied criticism of surveys is that they consume extensive resources which are therefore being denied to (and would go much further in) other forms of research. Roberts, for example, has stated: 'Sociologists are entitled to protest at this rampant and excessive fact-gathering' (Roberts, 1978, p. 28). But, since large-scale surveys tend to be conducted for policy rather than theoretical purposes, they do not generally compete for the same resources as other forms of leisure research. The considerable resources devoted to the conduct of particular policy-orientated leisure participation surveys would not be available for purely academic research purposes. In fact, virtually all academic use of such survey data is secondary and is often undertaken with little or no specific funding. Non-survey methods also have a place in policy research and this is being increasingly recognized by policy agencies; but such research tends to be conducted in addition to, rather than instead of, survey work.

The need for governments to base policy development and evaluation on quantitative statements about the whole community is likely to remain; it would therefore appear that leisure participation surveys are, in many countries, 'here to stay'. It would seem wise therefore for the leisure research community to make use of this resource and, where possible, to seek to influence the design of the surveys to maximize their utility for wider research purposes. Surveys have a role to play in leisure studies alongside other research methods.

THIS BOOK

The intention of the book is to establish as far as possible: how large scale national leisure participation surveys have been conducted in a number of economically developed countries; the extent to which they provide the basis for social analysis and policy development; the extent to which they might provide insights into the general nature and levels of participation in leisure activities of populations within participating countries; and whether it is possible to make some preliminary comparisons between countries.

This is an ambitious agenda, and the chapters of this book contain a wealth of material that goes a long way to answering these core questions. The contributions in the book vary enormously in terms of the social and economic characteristics of the countries involved and in terms of their experience with leisure surveys and leisure research generally. While some guidelines were given to authors concerning the structure of the chapters and issues to be discussed, these were indicative only. The result can be described as a 'structured diversity' approach (Ginsburg, 1992). Collectively they reveal a rich source of information and insights into leisure around the world. It is not proposed to summarize or evaluate the contributions here – but some conclusions are drawn in the final chapter.

NOTES

1. See ILO (Annual), OECD (Annual, 1993), UN (Annual).
2. Quotations from the Universal Declaration of Human Rights are from Brownlie, 1992.

REFERENCES

Aitchison, C. (1993) Comparing leisure in different worlds: uses and abuses of comparative analysis in leisure and tourism. In: Collins, M. (ed.) *Leisure in Industrial and Post-Industrial Societies*. Leisure Studies Association, Eastbourne, Sussex, pp. 385–400.

Appadurai, A. (1990) Disjuncture and difference in the global cultural economy. In: Featherstone, M. (ed.) *Global Culture, Nationalisation, Globalisation and Modernity*. Sage, London, pp. 295–310.

Bramham, P., Henry, I., Mommas, H. and Van Der Poel, H. (eds) (1993) *Leisure Policies in Europe*. CAB International, Wallingford.

Brownlie, I. (ed.) (1992) *Basic Documents on Human Rights.* Clarendon Press, Oxford.

Cichetti, C. (1972) A review of the empirical analyses that have been based upon the National Recreation Survey. *Journal of Leisure Research* 4, 90–107.

Clarke, J. and Critcher, C. (1985) *The Devil Makes Work: Leisure in Capitalist Britain.* Macmillan, London.

Dumazedier, J. (1982) *The Sociology of Leisure.* Elsevier, The Hague.

Featherstone, M. (1990) Global culture: an introduction. In: Featherstone, M. (ed.) *Global Culture: Nationalism, Globalisation and Modernity.* Sage, London, pp. 1–14.

Feldheim, P. and Javeau, C. (1977) Time budgets and industrialization. In: Szalai, A. and Petrella, R. (eds). *Cross-National Comparative Survey Research: Theory and Practice.* Pergamon, Oxford, pp. 201–230.

Filipcova, B. (ed.) (1972) Special issue on socialist life style. *Society and Leisure* No.3.

Ginsburg, N. (1992) *Divisions of Welfare: A Critical Introduction to Comparative Social Policy.* Sage, London.

Hantrais, L. and Kamphorst, T.J. (eds) (1987). *Trends in the Arts: A Multinational Perspective.* Giordano Bruno Amersfoot, Voorthuizen, Netherlands.

Hantrais, L. and Samuel, N. (1991) The state of the art in comparative studies in leisure. *Loisir et Société/Society and Leisure* 14 (2) 381–398.

Henderson, K. (1991) *Dimensions of Choice: A Qualitative Approach to Recreation, Parks, and Leisure Research.* Venture, State College, PA.

ILO (International Labour Office) (Annual) *Year Book of Labour Statistics.* ILO, Paris.

Jarvie, G. and Maguire, J. (1994) *Sport and Leisure in Social Thought.* Routledge, London.

Kamphorst, T.J. and Roberts, K (1989) *Trends in Sports: A Multinational Perspective.* Giordana Bruno Culemberg, Voorthuizen, Netherlands.

Kamphorst, T.J., Tibori, T.T. and Giljam, M.J. (1984) Quantitative and qualitative research: shall the twain ever meet? *World Leisure and Recreation* 26 (6), 25–27.

Kopytoff, I. (1987) The international African frontier: the making of African political culture. In: *The African Frontier.* Indiana University Press, Bloomington, p. 10.

Marans, R.W. and Mohai, P. (1991) Leisure resources, recreational activity, and quality of life. In: Driver, B.L. *et al.* (eds) *Benefits of Leisure.* Venture, State College, PA, pp. 351–364.

Mercer, D. (1994) Monitoring the spectator society: an overview of research and policy issues. In: Mercer, D. (ed.) *New Viewpoints in Australian Outdoor Recreation Research and Planning.* Hepper/Marriott Publications, Melbourne. pp. 1–28.

Novak, S. (1977) The strategy of cross-national survey research for the development of social theory. In: Szalai, A. and Petrella, R. (eds) *Cross-National Comparative Survey Research: Theory and Practice.* Pergamon, Oxford, pp. 3–48.

OECD (Organization for Economic Cooperation and Development) (Annual) *Tourism Policy and International Tourism in OECD Member Countries.* OECD, Paris.

OECD (Organization for Economic Cooperation and Development) (1993) *OECD Health Systems: Facts and Trends 1960–1991.* OECD, Paris.

Olszewska, A. and Roberts, K. (eds) (1989) *Leisure and Lifestyle: A Comparative Analysis of Free Time.* Studies in International Sociology, Vol. 38. Sponsored by International Sociological Association. Sage, London.

ORRRC (Outdoor Recreation Resources Review Commission) (1962) *National Recreation Survey,* Study Report No. 19, Government Printing Office, Washington DC.

Perez de Cuellar, J. (1987) Statement. *World Leisure and Recreation* 29 (1), 3.

Przeworski, A. and Teune, H. (1973) Equivalence in cross-national research. In: Warwick, D.P. and Osherson, S. (eds). *Comparative Research Methods*. Prentice-Hall, Englewood Cliffs, NJ, pp. 119–137.

Roberts, K. (1978) *Contemporary Society and the Growth of Leisure*. Longman, London.

Rojek, C. (1989) Leisure and recreation theory. In: Jackson, E.L. and Burton, T.L. (eds) *Understanding Leisure and Recreation: Mapping the Past and Charting the Future*. Venture, State College, PA, pp. 69–88.

Rojek, C. (1995) *Decentring Leisure: Rethinking Leisure Theory*. Sage, London.

Szalai, A. (ed.) (1972) *The Use of Time: Daily Activities of Urban and Suburban Populations in Twelve Countries*. Mouton, The Hague.

UN (United Nations) (Annual) *Statistical Yearbook*. UN, New York.

Warwick, D.P. and Osherson, S. (1973) Comparative analysis and the social sciences. In: Warwick, D.P. and Osherson, S. (eds) *Comparative Research Methods*. Prentice-Hall, Englewood Cliffs, NJ, pp. 3–41.

WTO (World Tourism Organization) (Annual) *Yearbook of Tourism Statistics*. WTO, Madrid.

2 Australia

S. DARCY AND A.J. VEAL

INTRODUCTION: EARLY SURVEYS

In Australia, as in many countries, national leisure participation surveys were preceded by local and regional surveys. It is, however, notable that these early efforts adopted a wide variety of techniques and approaches, so that their results are not comparable and trends cannot be measured. One local study, published under the simple title *Leisure*, was undertaken as early as 1958 for a religious charitable organization in an unidentified 'Australian housing estate' (Scott and U'Ren, 1962). Almost 200 respondents were asked how they usually spent their weekday evenings and Saturdays and Sundays and, setting the tone for all future surveys, the study report concluded:

> The most striking characteristic of the study was the concentration of leisure-time activities in and around the home. Home activities occupied most of the time of most people on Saturday mornings and afternoons, and Sunday mornings. It is difficult to know whether this preoccupation was by choice or necessity. However, most of the purely recreational and optional pursuits, such as reading and watching television, which absorbed a large amount of time, were also centred on the home.
>
> (Scott and U'Ren, 1962, p. 2)

The study noted that the domination of people's leisure time by home-centred and privatized leisure activities left a diminished role for provision by 'churches, municipal councils and voluntary associations'.

© 1996 CAB INTERNATIONAL. *World Leisure Participation* (eds G. Cushman, A.J. Veal and J. Zuzanek)

In 1974 the federal government's Department of Tourism and Recreation established a National Recreation Survey Task Force to 'define techniques for a national recreation survey and to prepare guidelines for the preparation of regional leisure plans'. The Task Force commissioned the *Gippsland Pilot Leisure Study* (Urbangroup, 1976) which, because of its methodological flaws, in fact provided little guidance for future research. The national Cities Commission's 1975 study, *Australians' Use of Time*, was based on a survey conducted in Melbourne and Albury-Wodonga, a developing city on the Victoria New South Wales border. With a total sample size of some 1500 adults, it required respondents to keep a diary of their activities for a 24-hour period. It was therefore primarily a 'time-budget' study rather than a 'participation' study and relates more directly to later time-budget studies (Mercer, 1985; ABS, 1988, 1994; Bittman, 1991) than to subsequent leisure participation surveys.

STATE, REGIONAL AND LOCAL SURVEYS

Australia has six states and two 'territories', and a number of surveys have been conducted at this level. The publication *Recreation, Wilderness and the Public* (McKenry, 1975) reported the results of a 1974 postal survey of adults in the state of Victoria, with an effective sample size of 1500. This did not serve as a model for later leisure surveys in that it was limited primarily to outdoor recreation and the main participation question referred to activities undertaken in the previous 12 months. The use of 12 months as a 'reference period' followed the practice of American surveys, but subsequent surveys in Australia have been designed in the belief that respondents' accuracy of recall does not extend over such a long time period and it is sounder practice to ask people about their more recent patterns of activity, for example, over the last week or month.

At regional level, the *Metropolitan Adelaide Recreation Study* of 1973/74 consisted of four quarterly household interview/diary surveys, returning to the same households for each survey (South Australia State Planning Authority, 1976). The combined sample size for the four surveys was a massive 27,000. The survey related to participation in away-from-home activities in a one-week period, but in this case, while it was not a time-budget study as such, respondents were asked to keep a diary of their activities, rather than being asked to recall activities from a checklist. The Adelaide study was followed by a series of state-wide surveys conducted in South Australia in 1984 (South Australia Department of Recreation and Sport, 1984). Based on participation in the previous week and on an extensive list of all types of leisure activity, these surveys provided the model for future national surveys.

Other surveys worthy of mention include the 1976 study of the Melbourne metropolitan area conducted in the form of the *Geelong Recreation Study* (John Paterson Urban Systems, 1977) which had a sample of 1800 interviews. Respondents were asked about participation in 36 outdoor recreation activities over the previous 12 months. The analysis reflected the preoccupation of many leisure researchers at that time with the construction of quantitative models of recreational choice. The independent Hunter Valley Research Foundation has been responsible for a number of surveys covering two or more local government areas in the New South Wales Hunter Valley Region (Hunter Valley Research Foundation, 1977, 1987). Numerous surveys were conducted at the local authority level in the 1970s, the 1980s and into the 1990s, over 100 having been identified nationwide (Veal, 1993a).

NATIONAL SURVEYS

The first national survey of leisure participation in Australia was the 1975 *Leisure Activities Away from Home*, which was part of the broader General Social Survey conducted by the Australian Bureau of Statistics (ABS, 1978). The survey involved a large sample of 8400 households containing 18,700 persons aged 15 years and over, each of whom supplied participation data. The main question asked was: 'In which activities do you spend most of your leisure time away from home?' A 'prompt' indicated that this was to include 'all activities for the last 12 months – that is for any season'. Space was provided for respondents to report up to seven activities. Extensive, expensive and competently conducted as this survey undoubtedly was, it did not provide a model for subsequent surveys, except in the list of individual activities which emerged. Asking people only about the activities on which they 'spend most time' has not been widely accepted as the best approach to gathering participation data, since the term 'spend most time' lacks precision and, further, some activities which may not in fact take up very much time – for example playing squash – may be significant to the participant. The 12 months reference period was also abandoned in later surveys.

A number of surveys have been conducted over the years on participation in *sport and physical recreation* rather than leisure as a whole, notably those initiated by the federal government Department of Sport, Recreation and Tourism[1] in 1984 and repeated in subsequent years (DASETT, 1988b). In summer and winter surveys samples of some 3500 adults were asked whether they had participated in 'physical activity', 'physical exercise' or 'physical recreation' in the previous 2 weeks, together with questions about constraints on participation. Since 1987 a market research company, Brian Sweeney and Associates, has conducted periodic surveys of 1500 adults in six state capital cities, asking respondents which of 40 sports they have either participated in (no time period specified), attended as spectator, watched on television or listened to on radio, together with questions on media use (Brian Sweeney Associates, 1989).

The most significant vehicle has been a series of National Recreation Participation Surveys (NRPS), conducted on behalf of the federal Department of the Arts, Sport, the Environment, Tourism and Territories (DASETT) between 1985 and 1991. The rest of this chapter therefore focuses on the NRPS series.

NATIONAL RECREATION PARTICIPATION SURVEY (NRPS)

The NRPS was conducted on behalf of the federal government by the market research company AGB:McNair. It was conducted on six occasions between Autumn 1985 and Summer 1991, on each occasion with a nationwide sample of over 2000 individuals aged 14 years and over (see Table 2.1). The surveys are the fulfilment of a policy of the federal government, as set out in a Ministerial statement in 1985:

> A first priority must be the creation of a comprehensive data base covering what Australians are doing in their leisure time, whether they are participating in their preferred activities or not, reasons why they may not be participating in preferred activities, and their attitude to recreation ... The Government will be working towards this end through the conduct of regular national recreation participation and attitudinal surveys which will take account of seasonal variations in recreation participation patterns and will provide a longitudinal perspective on recreation.
>
> (Brown, 1985, p. 26)

The NRPS has assumed some significance as a model and point of comparison for other surveys. It used a precise methodology, drawing on experience of surveys at state level, particularly in South Australia, as discussed above. Respondents were asked about participation in a specified list

Table 2.1. National Recreation Participation Survey details, Australia.

Year	Season	Sample size
1985	Autumn	2500
1985	Spring	2500
1986	Summer	2500
1986	Autumn	2500
1987	Spring	2068
1991	Summer	2103

of 91 activities over the week before interview, with seasonality being addressed by conducting surveys at different times of the year.

In addition to the provision of current information on participation the initial aims of the survey were to measure longitudinal trends and also to examine the benefits participants derive from recreation activities. A report on historical comparisons and benefits of recreation activities was to be published but to date has not appeared. Since the survey was conducted on a seasonal basis and only two comparable seasons' surveys have been replicated, assessment of longitudinal trends will not be possible, since the series has now been terminated.

The survey was administered in the form of an interviewer completed questionnaire. Respondents were asked to indicate activities they had participated in during their free time in the previous week. Show cards were utilized as *aides memoire*, showing the selected activities in five groups, namely: (i) home-based activities; (ii) social/cultural activities; (iii) organized sports; (iv) informal sports (in the tables presented here, organized and informal sport participation has been combined); (v) recreational activities. The 1991 survey also included information on participation in the previous month. In addition to the basic participation questions, respondents were also asked about frequency of participation in the week, leisure facilities visited and activities and facilities respondents would like to have participated in or used but had not, and reasons for this. In addition considerable sociodemographic detail was collected on respondents, including: age, gender, marital status, education level, country of birth, employment status and occupation, personal and household income, and size and composition of household.

The results of all the surveys have been published by DASETT and its predecessor department in individual reports (DSRT, 1986a, 1986b, 1986c, 1986d, 1986e; DASETT, 1988a, b, c, 1991), including tables and brief descriptive commentaries. There is no evidence publicly available of policy developments based on the survey findings. Some use of the data by state governments and local authorities in recreation planning and policy making is apparent (Veal, 1992). There has been some academic use of the data, but this could hardly be described as extensive. A collection of commentaries on the 1985/86 surveys, written by academics, was compiled and published in 1989 by DASETT (1989), and some independent use has been made of the data by academics for social analysis (McKay, 1990; Parker and Paddick, 1990; Veal, 1993b, 1993c; Darcy, 1993; Veal and Cushman, 1993) and for demand forecasting (Veal, 1987, 1988, 1991a, 1991b; Veal and Darcy, 1993).

The short, one-week, reference period of the NRPS should be noted before examining participation patterns in detail. The one-week period is useful for planning and management purposes, since leisure service and facility management is often geared to weekly cycles and the capacity of facilities can usefully be viewed in terms of 'visits per week'. The weekly measure is also useful for analysis of physical recreation, since, for participation in physical recreation to provide significant health benefits, it must take place at least weekly. However, for marketing purposes, for many activities the weekly figure is less useful. This applies particularly in the cultural area – for example very few visitors to museums or zoos visit weekly, so the weekly number of visitors is likely to include only a small proportion of the total 'market', although there is no reason to suppose that it is not representative of the total market in terms of socioeconomic characteristics.

A further feature of the NRPS data presented below should also be noted: that is, that the data presented relate to *summer*. For some activities, particularly winter sports, the data may therefore be misleading. Other versions of the NRPS provide seasonal data, but it is beyond the scope of this chapter to explore the seasonal dimension.

PARTICIPATION

Table 2.2 presents the basic participation data from the NRPS for summer 1991. The most popular activities, as in the 1950s study reported at the beginning of this chapter, were home based. Even among the social/cultural activities taking place outside the home, visiting friends and relatives (visiting *other people's* homes) was by far the most popular activity. While the physically undemanding activities of watching television/videos and 'relaxing, doing nothing' are the most popular home-based activities, the amount of physically active home-based activity is notable, with more than a third of respondents reporting engagement in *exercise and keep fit* activities, almost a quarter *swimming* in their own or friends' home swimming pools and 41% doing some *gardening*. A unique feature of the 1991 NRPS was the inclusion of *talking on the telephone* (for at least 15 minutes), which was engaged in by almost half the adult population.

Among the social/cultural activities, in addition to visiting friends and relatives referred to above, the most popular activities are *dining out* (32%) and *shopping for pleasure* (30%). The inclusion of the latter activity was an innovation of the 1991 NRPS and is rare in world terms (but see New Zealand chapter on the use of shopping centres). *Driving for pleasure* (19%), *church activities* (14%), *visiting pubs* (14%) and going on *picnics/barbecues* (14%) are the other popular activities in this group.

Among the sporting activities *swimming* is far and away the most popular, with 15% participation. *Tennis* comes a poor second, with only 6%, then *cycling* (5%) and *golf* (4%). These are the only sporting activities with 4% or more participants. *Football* is disadvantaged in this comparison, since it is a winter sport; however, the division of football into five codes in Australia, and

Table 2.2. Leisure participation, Australia, summer 1991.

% participating in week prior to interview			
Home-based		Sport	
Watch TV	93.6	Athletics	0.5
Listen to radio	77.1	Gymnastics	0.4
Entertain at home	35.8	Basketball	2.0
Electronic & comp. games	11.2	Netball – indoors	0.6
Exercise, keep fit	35.7	Netball – outdoors	0.8
Swim in own/friends' pool	23.3	Tennis	5.8
Play musical instrument	8.8	Squash	1.9
Arts, crafts	21.4	Badminton	0.2
Reading	70.4	Cricket – indoor	1.5
Listen music	65.1	Cricket – outdoor	3.9
Gardening for pleasure	41.3	Baseball/softball	0.6
Indoor games	17.7	Rugby League	0.2
Outdoor play with children	28.9	Rugby Union	0.1
Talk on telephone (15 mins +)	48.8	Australian Rules Football	0.3
Relax, do nothing	58.0	Soccer – outdoor	1.0
Social/Cultural		Soccer – indoor	0.3
Visit friends/relatives	62.8	Touch football	0.9
Dining, eating out	31.7	Martial arts	0.9
Dancing, discotheque	5.9	Motor sport	0.4
Visit pub	13.8	Archery/shooting	0.5
Visit (licensed) club	9.7	Hockey/lacrosse – indoor	0.3
Movies	8.0	Hockey/lacrosse – outdoor	0.3
Pop concerts	1.6	Cycling	5.0
Theatre	1.1	Golf	4.1
Music recital/opera	0.5	Swimming	15.5
Other live performances	0.8	Surfing/lifesaving	2.5
Special interest courses	1.6	Horse-riding	0.6
Church activities	13.6	Rink sports	0.3
Library activities	7.3	Lawn Bowls	1.5
Museums, galleries	1.6	10-pin bowling	1.5
Exhibitions	0.9	Recreation	
Arts crafts	3.9	Walk dog	14.2
Hobbies	11.2	Walk for pleasure	26.6
Picnic/barbecue away from home	13.7	Aerobics	5.4
Visit parks	10.3	Jogging/running	3.9
Horse races/trots/dog races	1.7	Bushwalking/hiking	2.3
Sport spectator	6.6	Skateboarding	0.3
Drive for pleasure	19.0	Shooting/hunting	0.3
Bird watching	2.6	Fishing	3.4
Play electronic games	4.0	Non-power water activities	1.3
Shopping for pleasure	29.6	Water activities – powered	0.9

Source: NRPS, based on Centre for Leisure and Tourism Studies, 1994.
Sample size (persons aged 14+): 2103.

the fact that it is overwhelmingly a male pastime, means that, even in winter, no one football code attracts more than 2 or 3% participation.

The 'recreation' group of activities overlaps considerably with sport, but is intended to encompass more informal physical recreation activities. The most popular of these activities is *walking*, with a dog (14%) or without a dog (27%). The so-called boom activities of the 1980s, *aerobics* and *jogging/running* attract only 5 and 4% respectively.

SOCIAL INEQUALITIES

Differences in patterns of leisure participation are examined below in relation to three variables: gender, age, and economic status and occupation. In the discussion below, only the 40 activities with at least 5% participation are included.

Gender

Table 2.3 compares participation levels between men and women, with activities where the levels of participation are significantly different being indicated by an asterisk. Exclusion of activities with less than 5% participation results in the exclusion of many sports, where men predominate. Of the 40 activities retained, for 11 there is no significant difference between male and female participation levels; while 15 have higher participation rates among women and 14 among men. This might suggest more equality between the sexes than anticipated, but it presents only a limited picture. In particular it is notable that, of the 15 activities where women's participation rates are higher than men's, seven are home based, illustrating the constraints on women as regards activities outside the home environment.

Age

Table 2.4 presents participation levels by age group. Two features of these data should be noted. First, for 29 of the 40 activities, the group aged 60 and over has the lowest participation rate (as marked by *). In many surveys, where the list of activities is dominated by sports and other physically demanding pastimes, this might be expected, but this is not the case with this list of activities. Of course the 60-plus age group includes the 'old old' who are likely to be less active, and it is still the case that older age groups are economically disadvantaged in Australia, thus limiting their leisure options. Nevertheless older people generally have large amounts of leisure time at their disposal, it is

Table 2.3. Leisure participation by gender, Australia, summer 1991.

	% participating in week prior to survey		
	Male	Female	Total
Sample size (persons aged 14+)	1042	1059	2102
Watch TV/videos	93.5	93.7	93.6
Entertaining	32.6	38.9*	35.8
Electronic/computer games	13.7*	8.8	11.2
Exercising/keep fit	36.4	35.1	35.7
Swim in home pool	23.9	22.6	23.3
Play music instrument	9.8*	7.7	8.8
Art/craft/hobby	15.6	27.1*	21.4
Reading	64.5	76.2*	70.4
Listening to radio	77.1	75.5	76.3
Gardening	39.0	43.6*	41.3
Phone friends (15 min +)	36.3	61.1*	48.8
Listen to music	63.3	66.9*	65.1
Indoor games	17.4	18.0	17.7
Play outdoors with children	25.6	32.1*	28.9
Relax/do nothing	57.9	58.0	58.0
Visit friends/relatives	59.5	66.0*	62.8
Dine/eat out	31.8	31.6	31.7
Dance/go to discotheque	7.5*	4.3	5.9
Visit pub	18.8*	8.9	13.8
Visit club	12.1*	7.5	9.7
Movies	8.3	7.6	8.0
Pleasure shopping	23.0	36.1*	29.6
Hobbies	11.4	11.1	11.2
Church activities	11.6	15.5*	13.6
Library activities	6.1	8.5*	7.3
Picnic/barbecue	13.8	13.6	13.7
Spectator at sport	7.6*	5.5	6.6
Drive for pleasure	20.3*	17.7	19.0
Visit parks	10.2	10.4	10.3
Electronic games	6.1*	1.9	4.0
Walking the dog	12.7	15.6*	14.2
Walk for pleasure	23.8	29.4*	26.6
Aerobics	2.2	8.6*	5.4
Cricket/outdoor	6.4*	1.5	3.9
Cycling	6.0*	4.1	5.0
Fishing	5.2*	1.6	3.4
Golf	6.3*	2.0	4.1
Running/jogging	5.3*	2.5	3.9
Swimming	13.2	17.8*	15.5
Tennis	6.9*	4.7	5.8

Source: NRPS, based on Centre for Leisure and Tourism Studies, 1994.
* Statistically significant difference, at 5% level of confidence.

widely believed that they are becoming more socially active, and many public programmes are directed at the elderly. These factors do not, however, appear to have overcome the traditional stereotype of a largely inactive elderly population.

The second feature of the age-related data lies at the other end of the age-range, namely the 14–19-year-old group. They are the most active group in 20 of the 40 activities. As a small age-group (11% of the adult population), this does not mean that this age-group dominates the activity in every case, but it does illustrate the 'life cycle' thesis, that, as people age and family and work responsibilities accumulate, leisure participation declines. The low participation rates for the elderly suggest that, once lost, leisure habits are not regained.

Economic Status and Occupation

Table 2.5 shows participation rates by economic status and occupation. The first four columns refer to people not in full-time paid employment. The *retired* category overlaps with the *elderly* group already discussed, and the *student* group overlaps with the 14–19 year group also already discussed. The *home duties* group includes primarily women, but separate data on women in all of the economic status categories would be needed to discuss this group fully.

The *unemployed group* are of special interest here, given that, at the time of the survey, unemployment was around 10% in Australia for the first time since the 1930s. Of the 40 activities, there are 19 where the participation rate for the unemployed was equal to or higher than the average (these are marked [†] in the table). This suggests that the unemployed do use some of their largely unwelcome free time for leisure purposes, but it is notable that in only one activity – *fishing* – are the unemployed the *most* active group. As with the elderly, this indicates that more than time is required to lead an active leisure life.

For those in full-time paid employment, social inequalities in leisure participation are illustrated quite starkly. The highest participation rates among the four occupational groups are indicated by a star (if rates are within 1% they are judged equal). In all cases except one (*indoor games*) the employed groups have higher participation rates than the average. For 23 of the 40 activities the highest or equal highest participation rate is among the managerial/professional group. The clerical/sales group is the second most active with 20 of the highest or equal highest rates. Supervisors have the highest participation rate in only one activity, *tennis*; the skilled manual workers in only four activities – *dancing, pub-going, sport spectating* and *fishing*. Unskilled manual workers have the highest rate only in *playing indoor games*

Table 2.4. Leisure participation by age, Australia, summer 1991.

	% Participating in week before interview by age						
	14–19	20–24	25–29	30–49	50–59	60+	Total
Sample size	226	233	209	776	246	411	2102
Watch TV/videos	98.5†	94.0	93.9	92.9	92.6	92.5*	93.6
Entertaining	33.7	42.6	44.9†	38.5	34.6	24.1*	35.8
Electron/comp. games	33.8†	11.7	12.1	12.2	1.5	2.2*	11.2
Exercising/keep fit	53.7†	47.4	42.8	37.2	23.5	20.2*	35.7
Swim in home pool	45.0†	24.2	23.1	26.8	16.1	8.4*	23.3
Play music instrument	17.7†	11.3	8.5	9.0	4.6	4.5*	8.8
Art/craft/hobby	22.8	14.4	17.3	22.2	32.2†	18.5	21.4
Reading	67.7	69.5	66.5	73.4†	68.5	69.8	70.4
Listening to radio	80.5	81.5†	74.7	77.6	74.6	70.3*	76.3
Gardening	11.2	20.9	37.3	48.0	54.3†	51.2	41.3
Phone friends (15 min +)	58.5†	56.6	50.8	49.8	47.3	37.1*	48.8
Listen to music	81.1†	73.3	74.6	64.0	57.6	53.2*	65.1
Indoor games	35.1†	17.6	15.4	19.0	11.4	10.7*	17.7
Play outdoors	25.5	27.1	38.4	39.4†	19.8	12.5*	28.9
Relax/do nothing	66.9	68.7†	58.6	57.0	48.3	54.2	58.0
Visit friends/relatives	71.8	75.5†	73.3	60.2	69.3	46.4*	62.8
Dine/eat out	34.6	38.2†	36.2	32.9	30.1	22.5*	31.7
Dance/discotheque	16.8†	16.2†	10.0	2.4	1.6	1.1*	5.9
Visit pub	14.8	31.0†	23.6	13.8	6.2	3.3*	13.8
Visit club	6.8	10.1	9.5	8.6	13.9†	10.8	9.7
Movies	22.6†	11.5	10.1	6.5	3.7	2.1*	8.0
Pleasure shopping	35.9†	33.6	27.5	30.2	30.0	23.3*	29.6
Hobbies	15.6†	11.5	11.4	10.7	14.7	7.5*	11.2
Church activities	12.9	7.5	8.4	13.9	16.1	17.8†	13.6
Library activities	4.5	4.9	5.5	8.8	5.6	9.3†	7.3
Picnic/barbecue	9.8	15.6	19.4†	17.8	11.4	5.7*	13.7
Spectator at sport	12.1†	7.4	6.8	7.8	3.4	2.6*	6.6
Drive for pleasure	15.8	24.2†	19.6	21.4	18.6	13.2*	19.0
Visit parks	6.9	10.9	8.5	15.4†	6.7	5.4*	10.3
Electronic games	12.9†	4.3	3.8	4.4	0.5	0.2*	4.0
Walking the dog	20.1†	11.7	8.8	16.1	17.6	9.4	14.2
Walk for pleasure	19.2	24.5	26.2	28.4†	27.1	28.5†	26.6
Aerobics	9.5	12.5†	9.2	4.6	1.4	1.2*	5.4
Cricket/outdoor	11.4†	6.9	6.1	3.2	1.1	0.2*	3.9
Cycling	8.6†	8.9†	5.6	5.5	1.3	2.0	5.0
Fishing	5.1†	4.6†	2.8	3.8	2.1	2.0*	3.4
Golf	2.7	3.4	2.7	5.4†	4.6†	3.3	4.1
Running/jogging	5.9	9.0†	6.8	4.2	0.0	0.0*	3.9
Swim/dive/water polo	24.0†	19.9	19.0	18.6	8.0	5.4*	15.5
Tennis	9.9†	8.6	4.2	6.9	5.3	0.9*	5.8

Source: NRPS, based on Centre for Leisure and Tourism Studies, 1994.
* 60+ age-group lowest participation rate. † Highest rate.

S. Darcy and A.J. Veal

Table 2.5. Leisure participation by occupation, Australia, summer 1991.

| | % participating in week prior to interview | | | | | | | | | |
| | Not in full-time paid employment | | | | In full-time paid employment | | | | | |
	Retired (n = 323)	Unemployed (n = 144)	F/T Student (aged 14+) (n = 175)	Home duties (n = 383)	Manager/ Prof. (n = 392)	Supervisor (n = 53)	Clerical sales (n = 270)	Skilled manual (n = 181)	Unskilled manual (n = 134)	Total (n = 2055)
Watch TV/Videos	91.8	95.7†	99.6	94.7	93.7*	87.0	94.3*	90.0	91.5	93.7
Entertaining	22.9	33.9	31.2	37.6	45.0	33.3	46.6*	32.9	25.6	35.8
Electron/comp games	2.2	13.1†	35.0	9.3	13.5*	5.4	10.5	5.1	9.3	11.1
Exercise/keep fit	21.5	36.0†	56.8	30.9	43.4*	30.4	40.3	32.4	29.6	35.6
Swim in home pool	7.0	22.0	48.5	18.1	28.6*	20.3	28.0*	26.2	18.6	23.3
Play music instrument	4.1	7.7	22.5	6.8	12.0*	2.2	7.1	4.5	9.1	8.6
Art/craft/hobby	19.8	14.9	26.2	27.7	25.7*	17.6	23.0	11.2	10.7	21.6
Reading	66.8	64.1	77.6	72.2	79.3*	57.3	75.2	57.8	58.5	70.4
Listening to radio	72.0	79.5†	78.9	72.5	80.0	73.2	82.9*	72.4	72.6	76.3
Gardening	49.5	25.6	12.1	50.6	46.2*	44.9	41.8	35.8	41.0	41.4
Phone friends	32.3	45.6	63.0	61.6	54.1	36.1	55.5*	36.3	30.1	48.8
Listen to music	51.1	62.2	77.5	64.2	73.8*	57.3	73.8*	62.3	52.0	65.1
Indoor games	10.6	20.6†	36.7	20.0	16.4	10.0	17.1	11.1	17.5*	17.7
Play outdoors	12.5	18.8	29.7	41.6	31.1	25.2	33.3*	24.5	29.9	28.6
Relax/do nothing	56.0	60.6†	65.2	57.4	55.9	58.0	65.9*	48.3	54.5	57.9
Visit friends/ relatives	45.7	57.3	72.6	67.6	64.6	57.6	72.1*	61.0	62.5	62.7

Dine/eat out	22.8	27.4	36.2	21.7	45.5*	30.1	39.1	30.2	23.6	31.4
Dance/go to disco	1.2	8.8†	11.7	2.8	6.1	3.4	8.7*	8.9*	4.8	5.8
Visit pub	4.8	17.6†	12.7	4.7	19.6	16.2	18.4	24.5*	16.3	13.7
Visit club	14.0	10.9†	3.6	5.5	10.7	7.3	13.6*	10.7	8.3	9.8
Movies/drive-ins	2.7	5.6	25.6	3.2	10.3*	4.9	10.9*	6.6	2.4	7.9
Pleasure shopping	21.5	31.4†	34.0	35.1	33.4	33.9	36.7*	16.8	16.7	29.7
Hobbies	8.4	12.3†	14.1	10.5	14.2*	12.0	10.0	10.7	10.2	11.3
Church activities	14.2	9.4	14.9	16.1	16.3*	13.2	10.7	8.4	11.2	13.5
Library activities	9.9	7.9†	7.1	9.2	8.5*	2.6	6.5	2.4	3.3	7.4
Picnic/barbecue	6.2	10.8	11.1	11.7	22.6*	18.5	15.8	12.6	13.4	13.7
Spectator at sport	2.4	5.4	13.8	3.9	6.3	6.1	8.8	11.7*	5.1	6.5
Drive for pleasure	13.8	18.1	17.9	16.8	23.8*	16.7	24.7*	16.8	16.7	18.9
Visit parks	7.5	13.0†	8.7	10.3	15.3*	6.6	11.4	7.7	4.2	10.3
Electronic games	0.7	4.9†	13.8	2.1	4.6*	3.6	3.4	2.3	4.1	3.9
Walking the dog	9.9	14.6†	19.1	12.3	15.9	11.8	18.3*	10.1	14.8	14.1
Walk for pleasure	30.4	28.0†	17.3	30.8	29.2*	12.3	28.3*	16.2	22.3	26.4
Aerobics	0.8	2.8	12.3	6.1	5.1	2.4	9.5*	3.4	5.8	5.5
Cricket/outdoor	0.2	5.0†	11.2	0.9	5.6	1.9	5.7*	3.6	5.1	4.0
Cycling	2.2	7.0†	8.9	3.0	6.9*	4.3	6.9*	3.2	6.0	5.1
Fishing	1.7	9.2†	4.6	1.0	2.2	6.5	3.2	9.2*	2.2	3.4
Golf	3.6	0.9	2.0	2.1	8.1*	2.2	4.6	5.3	3.0	4.1
Running/jogging	0.0	3.1	9.1	0.4	9.0*	2.6	4.0	2.3	3.3	3.8
Swim/dive/water polo	4.6	12.6	30.8	14.0	21.9*	14.8	21.2*	9.3	9.0	15.6
Tennis	1.1	4.9	12.2	3.7	8.9	14.0*	6.9	4.2	2.5	5.7

Source: NRPS, based on Centre for Leisure and Tourism Studies, 1994.
* Highest rates among those in full-time employment (1% differences or less ignored). † Unemployed equal to or higher than average.

and *playing electronic games*, while for most of the rest of the activities their participation rates are significantly lower than the average.

Summary

This analysis indicates wide variations in patterns of leisure participation among social groups in Australia. Gender, age and socioeconomic status clearly demarcate differences in patterns of leisure participation. Some of the differences may be due to taste differences, but it is also clear that social and economic constraints are also at work, presenting barriers to participation. The NRPS contains a wealth of untapped data which could be used to explore these issues further.

THE FUTURE

In 1993 the task of conducting national surveys was taken over by the Australian Bureau of Statistics, using their omnibus *Population Survey Monitor*, but results from this survey are only just becoming available at the time of writing (Corporate Concern, 1994). This survey has been conducted on a quarterly basis, using substantially larger samples than the NRPS. However, the changed vehicle and changed format of questionnaire means that continuity has been lost and it is unlikely that it will be possible to study trends in participation from the 1980s. Even if use of the *Population Survey Monitor* continues on a regular basis, trend analysis will not be possible until the end of the decade.

NOTES

1. The federal government department responsible for leisure/recreation has changed its name on numerous occasions over the last two decades. Beginning as the Department of Sport, Recreation and Tourism, it later became the Department of the Arts, Sport, the Environment, Tourism and Territories. First tourism and then the arts were removed, so that, at the time of writing, the department is the Department of the Environment, Sport and Territories.

REFERENCES

ABS – see Australian Bureau of Statistics
Australian Bureau of Statistics (1978) *General Social Survey: Leisure Activities Away from Home.* (Cat. No. 4104.0.) ABS, Canberra.

Australian Bureau of Statistics (1988) *Information Paper: Time Use Pilot Survey – Sydney, May–June, 1987.* (Cat. No. 4111.1.) ABS, Sydney.

Australian Bureau of Statistics (1994) *How Australians Use Their Time.* (Cat. No. 4153.0.) ABS, Canberra.

Bittman, M. (1991) *Juggling Time: How Australian Families Use Time.* Office of the Status of Women, Canberra.

Brian Sweeney and Associates (1989) *Australians and Sport 1989: The Third Annual Survey of Sporting Participation, Attendance and Media Habits.* (3 Vols.) Brian Sweeney & Associates, South Melbourne.

Brown, J. (Minister for Sport, Recreation and Tourism) (1985) *Towards the Development of a Commonwealth Policy on Recreation.* Department of Sport, Recreation and Tourism, Canberra.

Centre for Leisure and Tourism Studies (1994) *Analysis of the 1991, Department of the Arts, Sport, the Environment and Territories National Recreation Participation Survey,* using disks provided by Social Science Data Archive, ANU, University of Technology, Sydney.

Cities Commission (1975) *Australians' Use of Time.* Cities Commission, Melbourne.

Corporate Concern (1994) *Available Data and Sources for the Sport and Recreation Industry.* Report to the Statistical Working Group of the Sport and Recreation Ministers Council. Corporate Concern, Adelaide.

Darcy, S. (1993) Leisure participation in Australia: the monthly data. *Australian Journal of Leisure and Recreation* 4 (1), 26–32.

DASETT – see Department of the Arts, Sport, the Environment, Tourism and Territories

Department of the Arts, Sport, the Environment, Tourism and Territories (1988a) *Recreation Participation Survey: October/November 1987.* DASETT, Canberra.

Department of the Arts, Sport, the Environment, Tourism and Territories (1988b) *Physical Activity Levels of Australians.* AGPS, Canberra.

Department of the Arts, Sport, the Environment, Tourism and Territories (1989) *Ideas for Australian Recreation: Commentaries on the Recreation Participation Surveys.* AGPS Canberra.

Department of the Arts, Sport, the Environment, Tourism and Territories (1991) *Recreation Participation Survey: February 1991.* DASETT, Canberra.

Department of Sport, Recreation and Tourism (1986a) *Recreation Participation Survey: April/May 1985.* DSRT, Canberra.

Department of Sport, Recreation and Tourism (1986b) *Recreation Participation Survey: Oct/Nov 1985.* DSRT, Canberra.

Department of Sport, Recreation and Tourism (1986c) *Recreation Participation Survey: February 1986.* DSRT, Canberra.

Department of Sport, Recreation and Tourism (1986d) *Recreation Participation Survey: May 1986.* DSRT, Canberra.

Department of Sport, Recreation and Tourism (1986e) *Recreation Participation Survey: July 1986.* DSRT, Canberra.

DSRT – see Department of Sport, Recreation and Tourism.

John Paterson Urban Systems Pty Ltd (1977) *Geelong Recreation Study Phase 1.* Department of Youth, Sport and Recreation, Melbourne.

Hunter Valley Research Foundation (1977) *The Hunter Regional Health Plan*. NSW Department for Sport, Recreation and Tourism, Sydney.

Hunter Valley Research Foundation (1987) *The Recreation Demands and Needs of the Coalfields Community*. NSW Department for Sport, Recreation and Tourism, Sydney.

McKay, J. (1990) Sport, leisure and social inequality in Australia. In: Rowe, D. and Lawrence, G. (eds) *Sport and Leisure: Trends in Contemporary Popular Culture*. Harcourt, Brace, Jovanovich, Sydney, pp. 125–160.

McKenry, K. (1975) *Recreation, Wilderness and the Public: A Survey Report for the Department of Youth, Sport and Recreation*. Dept. of Youth, Sport and Recreation, Melbourne.

Mercer, D. (1985) Australians' time use in work, housework and leisure: changing profiles. *Australian and New Zealand Journal of Sociology* 21 (3), 371–394.

Parker, S. and Paddick, R. (1990) *Leisure in Australia*. Longman Cheshire, Melbourne.

Scott, D. and U'Ren, R. (1962) *Leisure: A Social Enquiry into Leisure Activities and Needs in an Australian Housing Estate*. F.W. Cheshire, Melbourne.

South Australia Department of Recreation and Sport (1984) *South Australian Leisure Activities Survey Report*. (4 Volumes) DRS, Adelaide.

South Australian State Planning Authority (1976) *Metropolitan Adelaide Recreation Study: Basic Data Report No. 3: Home Interview Survey of Demands for Recreation Facilities*. SASPA, Adelaide.

Urbangroup (1976) *Gippsland Pilot Leisure Study*. National Recreation Survey Task Force, Canberra.

Veal, A.J. (1987) Leisure and the future. *Recreation Australia* 7 (3), 1–3.

Veal, A.J. (1988) Future demand for outdoor recreation: planning implications. *Recreation Australia* 8 (3), 12–17.

Veal, A.J. (1991a) *Australian Leisure Futures: Projections of Expenditure and Participation, 1991–2001*. Publication 12. Centre for Leisure and Tourism Studies, University of Technology, Sydney.

Veal, A.J. (1991b) *National Parks and Recreation Demand: Current and Future National Park Visitation in the Sydney Metropolitan Area*. Publication 10. Centre for Leisure and Tourism Studies, University of Technology, Sydney.

Veal, A.J. (1992) *Local Leisure: The Collection and Use of Community Leisure Participation Survey Data by Local Authorities*. Report to DASETT. Centre for Leisure and Tourism Studies, University of Technology, Sydney.

Veal, A.J. (1993a) *Local Leisure: The Collection and Use of Recreation Participation Data by Local Authorities*. Report to the Department of the Arts, Sport, the Environment, Tourism and Territories, Centre for Leisure and Tourism Studies, University of Technology, Sydney.

Veal, A.J. (1993b) Leisure participation in Australia, 1985–91: a note on the data. *Australian Journal of Leisure and Recreation* 3 (1), 37–46.

Veal, A.J. (1993c) Leisure participation surveys in Australia. *ANZALS Leisure Research Series* 1, 197–210.

Veal, A.J. and Cushman, G. (1993) The new generation of leisure surveys – implications for research on everyday life. *Leisure and Society* 16 (1), 211–220.

Veal, A.J. and Darcy, S. (1993) Australian demographic change and leisure and tourism futures. In Boag, A. *et al.* (eds) *Proceedings of Australian and New Zealand Association for Leisure Studies Inaugural Conference.* ANZALS/Centre for Leisure Research, Griffith University, Brisbane, pp. 340–346.

3 Canada

JIRI ZUZANEK*

INTRODUCTION

In 1978, this author was asked to review Statistics Canada's programme of cultural and leisure statistics and comment on the state of Canada's leisure research in general. At the time, four studies conducted by Statistics Canada, the 1967–1972 Canadian Outdoor Recreation Demand Study (CORD), and a major Ontario survey of outdoor recreation were available for comparison and analysis. In 1988, on the occasion of the Fifth International Conference on Cultural Economics in Ottawa, another three major studies of leisure participation and time use had been added to the list of previously examined surveys (Zuzanek, 1988). Today we have at our disposal another four large surveys of leisure participation and the use of time, which lend themselves to comparative and trend analyses. What do the surveys, past and present, tell us about everyday life and free time in Canada? How does leisure and cultural participation vary across different social groups? How has the use of time and leisure participation changed over the period of last 20–25 years? Can we answer these questions on the basis of evidence provided by the existing surveys? Before addressing these questions, a chronology of relevant Canadian surveys is presented.

* The author would like to express thanks to his colleagues Terry Smith and Dr Bryan Smale for invaluable help in data analyses.
© 1996 CAB INTERNATIONAL. *World Leisure Participation* (eds G. Cushman. A.J. Veal and J. Zuzanek)

CHRONOLOGY OF SURVEYS

Surveys conducted in Canada can be divided into three groups: covering the period 1967–1978, when the emphasis was on leisure participation and time use; the 1980s, when attention turned to physically active leisure; and the 1990s, when the focus has been on cultural participation. These periods are discussed in turn below.

The Period 1967–1978: Surveys of Leisure Participation and Time Use

In the 1960s, and particularly the 1970s, in common with national statistical agencies in many countries, government bodies in Canada began collecting systematic information about the involvement of the population in non-working time, leisure and cultural activities. The 1970s were marked by a promising beginning of a broadly conceived and systematic study of leisure and cultural participation, spearheaded largely by the Education, Science and Culture Division of Statistics Canada, under the direction of Yvon Ferland. This was also a period when Canada first launched studies of the daily use of time, of which leisure forms an integral part. A conscious effort was underway to generate a comprehensive database which would allow the assessment of present and future trends in the leisure and cultural participation of Canadians.

In 1967–1972, Parks Canada conducted a series of large surveys of participation in outdoor recreational activities and park attendance, known as the CORD (Canadian Outdoor Recreation Demand) surveys. These surveys varied in scale and were designed to provide information about actual and potential demand for outdoor recreation in Canada. Technical information on this and subsequently discussed studies is shown in Table 3.1.

In 1971–1972, a team of researchers from Dalhousie University, under the direction of Andrew Harvey, conducted the *Halifax Time-Budget Survey*. This survey was designed along the lines of the Multi-National Time Budget Survey, conducted in mid-1960s under the auspices of UNESCO, and directed by Alexander Szalai (Szalai *et al.*, 1972). The study provided one of the first glimpses into the use of time by urban Canadians, and served as a benchmark for many future time-use studies.

In March 1972, Statistics Canada launched its first major survey of Canadians' leisure participation. The survey, *Leisure Study – Canada 1972* (Kirsh *et al.*, 1973), was conducted as a supplement to the monthly *Labour Force Survey*. Respondents were asked to answer questions on how often they participated in selected leisure activities from 1 January 1972 to the day of the survey. Activities included attending live theatre, opera, ballet, classical

Table 3.1. Overview of Canadian leisure participation and time-use surveys: 1967–1992.

No.	Year of survey	Title of survey	Agency	Geographic universe	Sample size	Sample age	Time referent	Measurement of participation
1.	1967, '68, '69, '72	Canadian Outdoor Recreation Demand Study (CORD)	Parks Canada	Canada	3,000 to 6,000	10+; 18+	per life-time	% participation
2.	1971–72	Halifax Time-Budget Study	Institute of Public Affairs, Dalhousie University	Halifax-Dartmouth area (N.S.)	2,141	18–64	day/week	minutes
3.	1972	Leisure Study – Canada 1972	Statistics Canada	Canada	62,789	14+	Jan.-March	% of participation frequency
4.	1973–74	Tourism and Recreation Planning Survey (TORPS)	TORP Commission of Ontario	Ontario	10,230	12+	12 months	% of participation time estimates
5.	1975	Leisure Study – Canada 1975	Statistics Canada	Canada	30,161	14+	12 months	% of participation frequency
6.	1976	Survey of Fitness, Physical Recreation and Sport	Statistics Canada	Canada	50,816	14+	12 months	% of participation frequency
7.	1978	Survey of Leisure Time Activities: Reading Habits	Statistics Canada	Canada	20,735	15+	12 months	% of participation time estimates
8.	1981	National Time Use Pilot Study	Statistics Canada	Canada: 14 urban centres	2,686	18+	day/week	minutes
9.	1981	Fitness and Lifestyle in Canada	Fitness Canada	Canada	23,400	12+	12 months	% of participation
10.	1983–84	Leisure and Cultural Participation of Urban Canadians	Research Group on Leisure & Cultural Development, University of Waterloo	Kitchener–Waterloo–Brantford area, Ontario	418	18+	12 months	% of participation frequency time estimates
11.	1986	General Social Survey: The Use of Time	Statistics Canada	Canada	9,946	15+	day/week	minutes
12.	1988	Well-Being of Canadians	Canadian Lifestyle & Fitness Research Inst.	Canada	2,778	12+	12 months	% of participation frequency
13.	1990–91	Canadian Arts Consumer Profile Survey	Department of Communications, Ottawa	Canada	11,106	16+	12 months	% of participation frequency
14.	1992	General Social Survey: The Use of Time	Statistics Canada	Canada	8,996	15+	day/week/ month/ 12 months	% of participation
15.	1992	Survey of Consumer Product Preferences	Printing Management Bureau (PMB)	Canada	12,000	12+	12 months	% of participation

music, museums, art galleries, movies, sports events, watching television, reading, hobbies, and participation in active sport and physical activities.

From May 1973 through April 1974, the Tourism and Outdoor Recreation Planning Commission of Ontario, conducted an extensive survey of participation in leisure and recreational activities in Ontario, the *Tourism and Outdoor Recreation Planning Study*, known as TORPS. Ontario residents aged 12 years and over, were asked questions about their participation in outdoor recreational activities, free time, leisure preferences and constraints and barriers to participation, as well as travel and tourism activities.

In October 1975, Statistics Canada administered a follow-up to the 1972 survey: *Leisure Study – Canada 1975*, again conducted as a supplement to the *Labour Force Survey*. Respondents aged 14 years and over were asked to report their participation in 45 selected leisure and cultural activities over the period of the last 12 months, as well as for the July–August and September–October periods.

The 1976 *Survey of Fitness, Physical Recreation and Sport* focused on participation in a wide variety of physical activities. Similar to Statistics Canada's previous leisure studies, it was administered as a supplement to the annual *Labour Force Survey*, and monitored frequency, intensity and location of, as well as motivations for and barriers to, leisure participation. Leisure activities included were similar to those in Statistics Canada's 1972 and 1975 studies.

In February 1978, Statistics Canada conducted yet another leisure related study, the *Survey of Leisure Time Activities: Reading Habits*, also administered as a supplement to the *Labour Force Survey*. It collected information about Canadians' participation in a broad spectrum of leisure and cultural activities, but paid particular attention to reading habits and preferences.

The 1980s: Physically Active Leisure and Time Use

The initial outburst of leisure and cultural surveys subsided in the 1980s. The research interest shifted to the use of time and participation in physically active leisure, and gravitated institutionally from the Education, Science and Culture Division to the General Social Survey Division of Statistics Canada, and Fitness and Amateur Sport Canada.

In 1981, Statistics Canada conducted its first *National Time Use Pilot Study*. Respondents from 14 urban centres across Canada were interviewed by telephone on their use of time. A number of traditional leisure and cultural participation questions were also included (Kinsley and Casserly, 1984).

Also in 1981, Fitness and Amateur Sport Canada conducted the survey *Fitness and Lifestyle in Canada* (Fitness and Amateur Sport, 1983), which examined fitness behaviour extensively, and included a battery of physical health and lifestyle measurements.

In 1983/84, the Research Group on Leisure and Cultural Development of the University of Waterloo was contracted by the federal Department of Communications to conduct the *Survey of Cultural Participation and Arts Related Attitudes of Urban Canadians*. An hour-long interview was administered to a random sample of household heads and their spouses from preselected urban census tracts representing lower, middle and upper-middle class neighbourhoods in the Kitchener–Waterloo–Brantford area of South Western Ontario. The survey aimed at collecting information in six areas: (i) patterns of leisure and cultural participation; (ii) familiarity with the 'cultural code' (i.e. levels of knowledge in the domain of arts and culture); (iii) value orientations and taste culture characteristics; (iv) mass media habits and preferences; (v) attitudes toward cultural policy issues; and (vi) 'mental' or 'semantic' images of culture and the arts.

Late in 1986, Statistics Canada included in its *General Social Survey* (GSS), for the first time, a full-scale time-use study. Telephone interviews were completed with adult members of randomly selected households. Respondents provided information on their time use for the 24 hours of a day from the previous week, including primary activity, total duration of each activity in minutes, where the activity took place, and with whom. Activities identified by the respondents were classified into one of 99 specific categories which were subsequently organized under general classes such as work for pay, domestic work, personal care and free time.

In 1988, the Canadian Fitness and Lifestyle Research Institute (Ottawa), conducted the survey *Well-Being of Canadians* (Canadian Fitness and Lifestyle Research Institute, 1991), sponsored by the Campbell Soup Co. It examined life cycle related changes in physically active leisure of persons aged 12 years and over. It was similar in design to the *1981 Canada Fitness Survey*, and used a multi-stage, cluster designed sample.

The 1990s: Leisure, Cultural Participation and Time Use

The early 1990s were marked by a revived interest in studies of cultural participation, continuing interest in time-use studies, and an emerging interest in leisure and culture on the part of commercial and marketing agencies. The federal Department of Communications, Statistics Canada's *General Social Survey*, and the marketing firm Printing Measurement Bureau (PMB), appeared as major vehicles of large-scale research in the area of leisure and cultural preferences, and the daily use of time. Three surveys stand out in the early 1990s as being of greatest interest: the 1990–1991 *Canadian Arts*

Consumer Profile Survey, the 1992 *General Social Survey,* and the 1992 PMB *Consumer Product Preferences Survey.*

In 1990–91 the federal Department of Communications (DOC) conducted the largest ever study of arts participation in Canada, the *Canadian Arts Consumer Profile Survey.* This study consisted of three components, the *General Public Survey,* the *Performing Arts Survey,* and the *Visual Arts Survey.* The *General Public Survey* is of greatest interest here. Persons aged 16 years and over were interviewed by telephone, with 5475 interviews completed in the first wave, and 5631 in the second wave. The survey included questions on participation in a variety of leisure and cultural pursuits, and on attitudes towards leisure and cultural participation.

Statistics Canada's 1991–92 *General Social Survey* (GSS), similar to the GSS in 1986, contained a time-use component. Telephone interviews were completed with individuals randomly selected from households across Canada and respondents provided information on their time use for the 24-hour period of the previous day.

During March–July and September–December of 1992, the Printing Measurement Bureau conducted the *Survey of Consumer Product Preferences.* The sample was based on a national household probability design, with the metropolitan areas of Toronto, Montreal and Vancouver being over sampled. The survey involved personal in-home interviews, and a self-completion questionnaire with questions about lifestyle preferences, and participation in selected leisure activities.

SOCIAL DIFFERENCES IN LEISURE PARTICIPATION AND THE USE OF TIME

Social differences in leisure behaviour have been traditionally studied from three different perspectives: as differences in the amounts and composition of *leisure spending;* the rates, diversity and frequency of *leisure participation;* and as differences in the amount and uses of *leisure time.* Authors using these measurements of leisure have often arrived at seemingly contradictory conclusions with regard to the distribution of leisure resources between different demographic and socioeconomic groups. The following analyses draw on a number of the above Canadian leisure participation and time-use data sources to discuss, in turn, gender, age, life cycle and socioeconomic status (SES) differences in access to, and uses of, leisure time.

Gender

Early Canadian studies of leisure participation (*Leisure Study – Canada*, 1975, *Kitchener–Waterloo–Brantford Survey*, 1984, – see Tables 3.2, 3.3 and 3.4), indicate that, for many activities, men's and women's rates of participation do not differ significantly. However, women tend to engage in cultural activities more than men, while the reverse is true of outdoor and sporting activities, where men's participation rates are usually higher than women's. Table 3.4 shows that men and women in the Kitchener–Waterloo–Brantford urban area engaged in approximately the same number (spectrum) of leisure activities, and reported similar intensity (weighted frequency) of leisure participation.

A similar pattern of gender differences emerges from leisure studies of the 1980s and the early 1990s. Tables 3.5 and 3.6, based on the results of the 1988 Campbell's *Survey of Well-Being of Canadians* and the 1992 *General Social*

Table 3.2. Participation in selected leisure activities by gender and education: Canada 1975.

| | Percentage participating in 12-months period | | | | |
| | | Gender | | Education | |
	Total	Men	Women	Elementary	University
Watching TV	96	96	96	92	95
Listening to records	63	62	63	25	81
Reading newspapers	75	75	74	42	91
Reading magazines	58	52	61	14	87
Reading books	55	45	61	14	83
Attending commercial movies	54	56	52	12	74
Attending art films/ciné-clubs	8	9	8	2	17
Attending popular concerts	17	17	17	3	25
Attending live theatre	13	11	16	2	38
Attending classical concerts	8	7	10	1	25
Visiting art galleries	17	16	19	2	45
Visiting science museums	11	12	11	1	28
Visiting other museums	17	16	17	3	38
Visiting historic site	24	23	24	5	47
Visiting libraries	27	24	30	2	57
Playing music	12	11	12	2	18
Engaging in arts (paint, sculpt)	12	11	12	1	25
Engaging in crafts	17	6	24	6	28
Hobbies	34	32	33	13	47

Source: Statistics Canada; Leisure Study – Canada 1975 (Table 3.1: 5).

42 — *Jiri Zuzanek*

Table 3.3. Participation in selected leisure activities by gender, age and education: Kitchener–Waterloo–Brantford Survey, Ontario, Canada, 1984.

		Gender		Age (years)			Education		
	Total	Men	Women	< 25	35–44	65+	Primary	Secondary	University
Watching TV	95.0	96.1	94.6	100.0	96.6	94.7	94.4	93.0	95.7
Listening to records and tapes	68.0	66.3	68.8	85.7	62.5	50.9	88.8	95.3	88.2
Reading books for pleasure	77.0	71.3	80.4	76.2	77.3	80.7	68.3	76.7	86.0
Visiting public library	69.0	65.2	72.5	78.6	78.4	54.4	55.9	74.4	84.9
Going to movies	68.0	67.4	67.9	95.2	77.3	45.6	49.1	74.4	87.1
Attending rock or pop concert	22.0	26.4	18.3	69.0	10.2	3.5	12.4	20.9	26.9
Attending a sporting event	47.0	55.1	41.7	61.9	55.7	22.8	34.2	51.2	65.6
Attending live theatre	46.0	42.7	49.2	40.5	47.7	42.1	26.1	53.5	68.8
Attending opera or ballet	18.0	15.7	19.6	7.1	18.2	17.5	10.6	14.0	32.3
Attending classical music concert	28.0	25.8	30.0	23.8	28.4	29.8	19.3	25.6	47.3
Attending jazz concert	10.0	9.6	9.6	14.3	4.5	3.5	6.2	1.2	20.4
Visiting art gallery/art museum	35.0	36.0	35.0	31.0	34.1	40.4	23.6	32.6	59.1
Visiting historic site/museum	58.0	57.3	57.9	52.4	60.2	54.4	46.0	64.0	71.0
Visiting a fair or festival	68.0	68.5	67.5	83.3	79.5	47.4	55.3	73.3	76.3
Travelling overseas	17.0	19.1	15.8	9.5	17.0	26.3	11.2	17.4	28.0
Visiting national/provincial park	66.0	62.9	68.3	78.6	70.5	45.6	58.4	72.1	72.0
Camping/visit private cottage	77.0	77.5	76.7	85.7	79.5	52.6	68.3	80.2	82.8
Hiking/backpacking	46.0	48.9	42.9	66.7	58.0	14.0	37.9	44.2	62.4
Fishing/hunting	55.0	60.7	50.0	76.2	59.1	31.6	55.9	54.7	55.9
Sailing/canoeing/powerboating	55.0	56.2	54.2	78.6	65.9	26.3	47.2	53.5	67.7
Walking for exercise	93.0	92.1	93.8	92.9	95.5	89.5	90.7	93.0	97.8
Swimming	76.0	77.0	75.4	97.6	79.5	43.9	62.1	82.6	89.2
Jogging, calisthenics	59.0	61.8	57.5	85.7	63.6	28.1	44.1	68.6	76.3

Percentage participating in a 12-months period

Activity									
Skating, skiing	55.0	55.1	55.0	83.3	72.7	12.3	46.6	55.8	72.0
Tennis, racquet sports	41.4	42.7	40.4	73.8	53.4	8.8	31.7	43.0	53.8
Golfing	45.7	60.7	34.6	59.5	55.7	26.3	42.9	47.7	54.8
Curling	30.1	34.3	27.1	52.4	39.8	15.8	28.6	34.9	33.3
Bowling	51.2	48.3	53.3	66.7	60.2	22.8	54.7	46.5	52.7
Team sports (hockey, softball)	40.9	47.2	36.3	76.2	51.1	10.5	36.0	44.2	47.3
Strength sports (weights, box)	36.1	46.1	28.8	76.2	44.3	12.3	32.9	38.4	43.0
Play musical instrument/sing	52.6	48.3	55.8	71.4	55.7	29.8	43.5	54.7	62.4
Painting, sculpturing, pottery	34.9	34.3	35.4	50.0	47.7	15.8	32.3	36.0	43.0
Weaving, knitting, crocheting	56.5	29.2	76.7	52.4	64.8	35.1	57.8	59.3	47.3
Woodwork, carpentry	55.5	72.5	42.9	61.9	61.4	52.6	55.3	47.7	65.6
Gardening	80.6	76.4	83.8	61.9	84.1	82.5	79.5	86.0	77.4
Collecting (stamps, coins)	47.4	48.9	46.3	54.8	62.5	31.6	46.6	39.5	57.0
Gourmet cooking	70.1	56.7	80.0	66.7	78.4	42.1	68.3	67.4	75.3
Bingo	38.8	34.3	42.1	54.8	46.6	26.3	44.1	40.7	36.6
Playing cards, board games	84.7	84.8	84.6	92.9	83.0	64.9	79.5	88.4	89.2
Attending auctions	54.3	53.4	55.0	59.5	56.8	28.1	51.6	51.2	50.5
Pinball, pool, shuffleboard	46.4	51.1	42.9	71.4	55.7	15.8	42.9	51.2	50.5
Playing chess, checkers	49.3	54.5	45.4	59.5	58.0	19.3	43.5	47.7	58.1
Computer/video games	45.2	47.2	43.8	76.2	63.6	12.3	36.0	50.0	53.8
Dining out	91.9	88.8	94.2	95.2	92.0	80.7	88.2	94.2	97.8
Entertaining at home	87.6	86.0	88.8	71.4	92.0	91.2	85.7	84.9	93.5
Going to nightclub, disco	51.9	51.7	52.1	81.0	51.1	21.1	44.1	61.6	51.6
Working for community group	46.4	48.3	45.0	54.8	62.5	35.1	37.3	46.5	62.4
Working for a church group	52.9	50.6	54.6	50.0	61.4	45.6	49.7	52.3	61.3
Member of social/sports club	48.6	50.0	47.5	69.0	59.1	33.3	42.2	50.0	63.4
General interest courses	43.5	40.4	45.8	64.3	59.1	15.8	32.9	44.2	58.1

Source: Research Group on Leisure and Cultural Development, University of Waterloo (Table 3.1: 10).

Survey, point to gender differences in selected cultural, spectating, hobby-like and physically active pursuits. Activities such as attending baseball and football, playing hockey, tennis, basketball, baseball and football and golf, weight training, fishing, and woodworking were found to be more popular among men than among women, while attending live theatre and ballet performances, going to classical music concerts, walking for exercise, home exercise, aerobics, and social dancing attracted more women participants than men.

In general, studies of leisure participation over the past 20 years reveal that gender differences in leisure participation can be characterized as differences of *preference* rather than magnitude, diversity or intensity of leisure participation. If anything, women appear to engage in a more diverse spectrum of leisure pursuits than men.

The time-budget surveys, on the other hand, reveal a rather different situation. In particular, employed married women with children living at home, appear to be disadvantaged in their access to leisure time, compared

Table 3.4. Diversity and intensity of leisure participation by gender, age, education and income: Kitchener–Waterloo–Brantford, Ontario, Canada, 1984 (Composite scores: 1–100).

	Spectrum of participation*	Intensity of participation†
Total sample	53.2	33.3
Gender		
Men	53.5	33.1
Women	53.0	33.5
Age		
< 25	65.1	39.8
25–34	60.2	37.0
35–44	59.2	35.9
45–54	50.9	31.6
55–64	42.8	28.3
65+	36.7	24.8
Education		
Elementary	47.2	29.4
High School	54.4	34.0
University	62.2	39.0
Income ($C p.a.)		
< 16,000	48.2	29.8
16,000–19,999	56.7	35.1
20,000–29,999	57.3	35.1
30,000–39,999	52.1	34.3
40,000 +	61.4	37.7

* Percentage of leisure activities reported by respondent.
† Percentage of leisure activities weighted by frequency of participation.
Source: Leisure Studies Data Bank, University of Waterloo (Table 3.1: 10).

with men. The 1971–72 time-budget study in Halifax, showed clear male–female differences in their access to free time along employment lines. Employed women reported having an average of 4.0 hours of free time per day, compared to men's 4.9 hours, and 5.4 hours for the housewives (Elliot *et al.*, 1973). According to a 1971 time-budget study in Greater Vancouver, employed married women with children, had only 1.8 hours of free time per day, compared with 3.4 hours for the analogous category of men (Meissner *et al.*, 1975).

These findings are corroborated by the national time-use surveys of the 1980s and early 1990s. Data from the 1981 National Time-Budget Pilot Survey indicate that employed married men aged 19–34 with children under the age of five, had 273 minutes of free time per day, compared with 201 minutes for women in the same employment and family status category (Table 3.7). According to the 1986 GSS Time Use Survey, employed married men aged 25–44 with a child, had 249 minutes of free time per day, compared to 224 minutes for women in the same employment/family status category (Table 3.8). In 1992, the gap between men's and women's access to free time

Table 3.5. Participation in selected physically active leisure pursuits, Canada, 1981 and 1988.

	1981 Survey	1988 Survey		
	Total	Total	Men	Women
Walking for exercise	59.8	67.9	59.7	75.3
Bicycling	42.0	44.5	46.1	43.0
Jogging or running	29.6	20.0	24.6	16.0
Home exercise	31.0	33.6	29.0	37.8
Aerobics	10.3	15.1	7.9	21.6
Ice skating	23.5	24.1	26.6	21.9
Cross-country skiing	20.1	17.3	17.3	17.3
Downhill skiing	12.7	20.5	23.9	17.4
Ice hockey	10.4	10.0	19.4	1.6
Swimming	40.4	46.1	45.0	47.1
Gardening	33.3	55.9	58.5	53.6
Golf	13.1	19.4	28.1	11.7
Tennis	16.9	13.5	17.1	10.3
Weight training	6.2	13.3	19.0	8.2
Baseball, softball	12.3	18.5	25.5	12.2
Social dancing	13.9	36.9	31.3	42.0
Fishing	6.1	5.0	7.8	2.4
Volleyball	5.2	4.3	3.8	4.8
Bowling	8.7	18.8	18.9	18.8

Values are percentage participation for a 12-month period.
Source: Canadian Fitness & Lifestyle Research Institute; Leisure Studies Data Bank, University of Waterloo (Table 3.1: 9 & 12).

Jiri Zuzanek

Table 3.6 Participation (%) in selected leisure activities by gender, age and education: General Social Survey, Canada, 1992.

		Gender		Age (years)						Education		
	Total	Men	Women	<25	25–34	35–44	45–54	55–64	65+	Primary	Secondary	University
Watching film on VCR	70.4	73.4	67.9	92.3	85.9	81.5	64.9	47.7	27.3	64.5	75.2	81.8
Listening to records, CDs	81.5	83.1	80.3	95.9	90.0	85.8	79.9	70.0	56.7	70.6	84.5	90.4
Reading book as leisure	68.0	59.6	74.7	72.9	67.5	71.5	68.7	62.5	62.1	53.0	67.6	84.3
Borrowing book from library	33.2	28.9	36.7	48.2	34.1	40.2	37.6	21.2	19.1	20.5	30.0	53.1
Going to movies	48.0	50.4	46.1	79.9	38.4	52.4	39.9	23.0	15.4	29.5	47.4	69.9
Attending pop/rock concert	15.7	18.2	13.7	34.4	21.3	14.4	8.6	3.8	2.7	66.9	64.4	67.9
Attending professional sporting event	29.5	39.5	21.4	39.8	35.8	34.9	26.1	20.1	10.7	18.1	30.4	42.9
Attending drama	8.4	7.4	9.2	9.3	7.5	10.7	9.0	7.2	6.2	3.1	5.0	17.9
Attending comedy	9.5	9.1	9.8	7.5	9.2	11.2	12.6	10.1	6.7	4.8	8.2	16.0
Attending musical	10.4	8.9	11.6	9.1	9.4	12.7	12.7	12.9	6.9	43.6	49.1	48.6
Attending opera	3.6	3.0	4.0	2.7	3.0	3.8	5.0	4.7	3.2	33.3	28.1	33.4
Attending ballet	3.0	1.9	3.8	3.5	2.7	3.3	3.4	2.2	2.5	61.4	63.0	63.3
Attending symphonic music	4.6	4.2	4.9	2.4	3.1	5.9	6.8	6.0	4.9	30.0	34.2	47.5
Attending symphonic pops concert	2.3	2.1	2.4	1.2	1.5	2.9	3.2	3.3	2.3	16.8	24.0	21.7
Attending jazz/blues performance	5.7	6.8	4.9	7.8	7.9	7.3	5.0	2.2	1.0	15.1	18.2	31.4
Visiting museum or art gallery	33.7	33.0	34.2	32.1	33.1	41.7	37.2	32.0	23.9	18.7	28.3	54.2
Visiting public art gallery	20.0	18.4	21.2	19.1	18.7	25.5	23.0	18.9	13.8	49.8	50.6	67.3
Visiting historic site	28.1	31.0	25.8	27.0	27.8	36.5	29.4	28.0	21.9	16.9	24.5	42.8
Visiting a festival or fair	52.1	54.2	50.4	62.1	59.3	58.8	51.3	41.5	28.9	39.1	50.8	65.2
Playing musical instrument	18.1	19.9	16.7	28.6	17.3	18.4	17.6	13.9	11.3	13.9	15.3	25.9
Doing any crafts	33.8	15.7	48.4	24.9	32.0	30.3	36.8	40.5	43.7	31.9	33.1	32.8

Source: Statistics Canada; Research Group on Leisure and Cultural Development, University of Waterloo (Table 3.1: 14).

for employed respondents, aged 25–44 with children under five at home, widened to 44 minutes per day, with men having 243 minutes of free time per day compared with women's 199 minutes (Table 3.8).

These findings are paralleled by data from the 1992 survey about men's and women's subjective perception of being 'rushed' or pressed for time. The mean score of perceived time pressure on a 24-point scale was 14.8 for employed mothers aged 25–44, and 12.2 for employed women aged 35–64 with no children at home, compared with 12.4 and 10.7 for men in identical employment and family status categories (Table 3.9).

In sum, unlike leisure *participation*, gender differences as measured by men's and women's access to *free time*, are considerable. Studies by Shaw (1985), and Zuzanek and Smale (1992) demonstrate that gender differences in the distribution of free time are the strongest in middle age groups and

Table 3.7. Life cycle variations in the use of time (minutes per day): Canada, 1981.

	Work for pay	Domestic work	Sleep	Free time	TV
Men					
University student, single, 19–24	9	50	512	332	79.1
Employed, single, 19–24	349	86	480	346	97.6
Employed, married, child < 5, 19–34	379	159	457	273	126
Employed, married, school age child, 25–44	401	112	465	265	119
Employed, married, empty nest, 35–64	341	12	463	330	132
Retired, married, 65+	24	192	497	478	221
Retired, widower, 65+	0	274	485	470	258
Mean	240	140	1484	421	128
Women					
University student, single, 19-24	74	120	497	235	36
Employed, single, 19–24	358	104	467	301	66
Employed, married, child < 5, 19–34	320	290	495	201	78
Homemaker, married, child < 5, 19–34	21	439	487	308	110
Employed, married, school age child, 25–44	355	218	471	224	71
Homemaker, married, school age child, 25–44	13	401	465	341	103
Employed, married, empty nest, 35–64	297	211	451	266	70
Homemaker, married, empty nest, 35–64	17	303	495	410	127
Retired, married, 55+	0	263	531	373	141
Retired, widow, 55+	1	254	508	373	179
Mean	130	255	499	402	104
Total	181	203	492	411	115

Source: Statistics Canada; Leisure Studies Data Bank, University of Waterloo (Table 3.1: 8).

Table 3.8. Changes in daily free time (minutes per day) between 1986 and 1992 for selected life cycle groups: Canada.

Life Cycle Grouping Description	1986			1992		
	n	Mean	s.d.	n	Mean	s.d
Single, working men under 35 years of age	492	337	218	473	358	234
Married working men, 25–44, children under 5 years at home	527	249	184	437	243	180
Married working men, 35–64, no children under 5 years at home	382	301	199	398	310	190
Retired married men, 65 years of age or older	336	549	216	247	535	184
Single, working women under 35 years of age	408	302	195	328	307	199
Married working women, 25–44, children under 5 years at home	211	224	165	197	199	168
Married women working at home, 25–44, children under 5 years at home	370	299	157	292	269	167
Married working women, 35–64, no children under 5 years at home	245	281	185	312	255	170
Married women working at home, 65 years of age or older	136	430	159	154	449	150

Source: GSS, Statistics Canada; Research Group on Leisure and Cultural Development, University of Waterloo (Table 3.1: 8, 14).

families with a large number of children, and are more pronounced on weekends than on workdays.

Age and Life Cycle

Evidence available from the 1970s suggests that for both men and women, the mean number of leisure activities engaged in tends to decline with age. *TORPS* and *Leisure Study – Canada 1972*, indicate that participation rates in most leisure and cultural activities are highest among the 18–25 year old group, and lowest among retirees. According to *Leisure Study – Canada 1972*, the highest rates of participation for all leisure activities, with the exception of opera attendance, were reported by the age groups of 14–19 and 20–24, and the lowest by the age group 65+ (Table 3.10). According to *TORPS*, the mean number of activities participated in by those under 59 years of age was 13.8 compared with only 5.3 for those aged 60+. These findings seem to be in line with the 'disengagement' argument about gradual decline of activity levels among the elderly.

The 1984 *Kitchener–Waterloo–Brantford Survey* lends further support to some of these findings, yet suggests that participation in different leisure and cultural activities 'slopes' unevenly with age. According to Table 3.3, activities such as watching television, reading books for pleasure, woodwork and carpentry, playing cards and board games, entertaining at home and dining out, are relatively evenly distributed across the life cycle. Participation in many sporting and social activities such as team sports, jogging, skating, skiing, swimming, curling, and social or sport clubs, on the other hand, shows a clear pattern of decline over the life span. However, for a large group of spectator, attendance and recreational activities, such as attending movies and sporting events, visiting fairs, festivals, national and provincial parks, bowling, listening to records and tapes, painting and sculpting, and taking general interest courses, a discernable decline in participation occurs only after middle age. Another group of mostly cultural and hobby activities, such as visiting art galleries and museums, attending concerts of classical music and opera, visiting historic sites, and travelling overseas, shows an increased level of participation in mid-life cycle with a moderate decline occurring only in the post-retirement age.

These findings are, in general, supported by subsequent surveys of leisure

Table 3.9. Perceptions of being 'time rushed' and 'time pressured' by selected life cycle groups: Canada, 1992.

Life cycle group description	Time Pressure Index*			Do you feel more rushed than five years ago?		
	n	Mean	s.d.	Less	Same	More
Single, working men under 35 years of age	480	9.9	6.4	18.7	20.7	60.6
Married working men, 25–44, children under 5 years at home	447	12.4	6.8	14.2	23.3	62.5
Married working men, 35–64, no children under 5 years at home	881	10.7	6.9	19.7	35.0	45.3
Retired married men, 65 years of age or older	247	3.4	3.5	50.4	41.5	8.1
Single, working women under 35 years of age	359	10.7	6.7	11.5	21.4	67.0
Married working women, 25–44, children under 5 years at home	183	14.8	6.9	5.9	9.6	84.6
Married women working at home, 25–44, children under 5 years at home	299	11.5	6.8	22.2	13.7	64.1
Married working women, 35–64, no children under 5 years at home	663	12.2	7.2	15.6	30.0	54.3
Married women working at home, 65 years of age or older	149	4.9	5.0	46.7	38.8	14.5

* Based on a 5-item computed measure ranging from 'very low' (value = 1) to 'very high' (value = 24) pressure.
Source: GSS, Statistics Canada; Research Group on Leisure and Cultural Development, University of Waterloo (Table 3.1: 14).

Table 3.10. Participation (%) in selected leisure and cultural activities by age: Canada, Jan–March, 1972.

	Total	14–19	20–24	25–34	35–44	45–64	65+
Watching TV	94	96	95	96	95	94	89
Listening to records	50	77	66	55	47	36	18
Attending movies	35	59	60	43	31	20	7
Attending non-classical music concert	13	32	22	12	8	4	2
Attending sports events	23	43	31	25	23	14	4
Attending live theatre	17	13	9	8	8	8	4
Attending opera or operetta	1	2	1	1	2	2	1
Attending ballet performance	1	1	1	1	1	1	0
Attending classical music concert	6	9	6	5	6	6	3
Visiting public art gallery*	6	10	9	6	5	4	2
Visiting museum*	7	11	9	7	7	5	2
Visiting historic site*	10	14	15	10	9	9	3
Visiting fair, exhibition, carnival	11	19	14	12	11	8	3
Engaging in arts, crafts or music	25	33	25	27	23	21	16
Hobbies	12	17	15	13	10	8	6
Participating in sports activities	23	48	33	26	20	12	4
Participating in physical activities	20	43	27	20	15	10	5

Source: Statistics Canada; Leisure Study – Canada 1972 (Table 3.1: 3).
* Sum of free and paid attendance.

and cultural participation, including the *1990–91 Canadian Arts Consumer Profile Survey*, and the *1992 General Social Survey* of participation in selected leisure and recreational activities (see Table 3.6).

If the ageing pattern for *leisure participation* can be, for the most part, characterized as that of a 'calibrated decline', the relationship between age and access to *free time* resembles a 'bi-polar curve', with the largest amounts of free time being reported by the youngest and the oldest age groups, and the lowest amounts by middle aged groups. According to *TORPS* (1978), Ontario respondents in the age group 70+ reported having 6.3 hours of free time per day, compared with 4.3 hours among the 35–44-year-old group, and 5.3 hours in the age group 20–24.

It has been suggested that age groupings often serve as substitutes for *life cycle* situations characterized by changing constellations of social roles. The 'time crunch' of the middle age should be attributed to a cumulative pressure of multiple career, employment, family and status roles rather than biological age *per se* (Zuzanek, 1979; Zuzanek and Box, 1988). This multiple role strain is particularly obvious in the case of employed married women and men with small children at home.

According to the data from the 1986 General Social Surveys (Table 3.8), single employed men aged 35 or less reported having 337 minutes of free time per day, compared with 249 minutes for men aged 25–44, with small children at home, and 549 minutes for retired married men aged 65 and more. Likewise, employed single women under the age of 35, reported having 302 minutes of free time per day, compared with 224 minutes for employed married women aged 25–44 with small children at home, and 430 minutes for married women aged 65+ at home. Similar findings are reported by the *1992 General Social Survey*. Analyses of the subjective perceptions of time pressure among different age and life cycle groups also indicate that middle aged employed respondents with children at home experience greater time pressure than younger and retired respondents (Table 3.9).

Educational and Occupational Status

A number of authors (Wilensky, 1963; Ennis, 1968; Wippler, 1970; Cheek and Burch; 1976, Zuzanek, 1978) have examined differences in leisure behaviour associated with social and economic status. According to Wippler (1970), socioeconomic status, in particular education, represents an important determinant of the use of time and of leisure behaviour. The findings of Canadian leisure participation surveys from the 1970s and 1980s (Tables 3.2, 3.3, 3.4, 3.6, 3.11) indicate that levels of leisure and cultural participation are, indeed, strongly affected by respondents' socioeconomic status.

Table 3.12, based on the findings of the *Leisure Study – Canada 1972*, indicates that higher educational status is associated with higher levels of participation in most leisure activities, particularly cultural and physically active ones. Higher occupational status also predicates more active leisure and cultural participation, although the relationship is not as 'clear-cut' as with education. Compared with blue collar workers, white collar employees report higher rates of participation in most leisure activities. Professionals report higher levels of cultural participation than white collar employees, but 'lag' behind white collar employees in physically active leisure pursuits and spectator activities such as attending movies and sporting events.

Similar findings are reported in the *Leisure Study – Canada 1975*; the 1984 *Survey of Leisure and Cultural Participation in Kitchener–Waterloo–Brantford* area, and the 1988 *Canada Fitness Survey*. According to the *Leisure Study – Canada 1975*, university graduates reported higher levels of participation than respondents with elementary education in all surveyed activities (Table 3.2). Table 3.3 shows that, in the Kitchener–Waterloo–Brantford area in 1984, higher educational and income status was associated with greater diversity and intensity of leisure participation. The few activities for which higher educational status does not signify an increase in participation, include weaving, knitting, crocheting, bingo, attending auctions, baseball, softball, and possibly fishing and social dancing (Tables 3.3 and 3.11).

As in the case of gender and age, the relationship between leisure and socioeconomic status takes a different form when viewed through the lens of *free time* rather than from the perspective of diversity and intensity of *leisure participation*: greater access to free time does not always go hand in hand with a greater interest in and diversity of leisure participation. The socioeconomic status groups reporting highest levels of leisure and cultural participation are often also those who are most pressed for time and least 'leisurely'.

According to the 1986 *General Social Survey* (Statistics Canada, 1990), respondents with the highest socioeconomic status (SES) (measured on the Blishen's scale of occupational prestige (Blishen *et al.*, 1987)), reported having 244 minutes of free time per day, compared with 262 minutes for the group with medium SES, and 278 minutes for the group with low SES (Table 3.13). A similar trend emerges when the amounts of free time are compared for managers (252 minutes per day); professionals (259 minutes); clerical employees (268 minutes); and blue collar workers (272 minutes).

Table 3.11. Participation in selected physically active leisure pursuits by education, Canada, 1988.

	Less than high school	High school	Some post high school	University
Walking for exercise	62.0	63.6	70.8	78.4
Bicycling	51.7	42.4	42.0	50.1
Jogging or running	25.9	17.8	19.1	26.2
Home exercise	30.2	31.6	33.7	38.0
Aerobics	12.5	13.1	16.3	19.3
Ice skating	33.7	22.3	20.6	25.8
Cross-country skiing	15.6	14.4	13.0	27.6
Downhill skiing	18.6	18.9	19.5	27.0
Ice hockey	12.3	11.2	7.0	7.7
Swimming	47.9	43.9	44.2	51.1
Gardening	44.1	53.3	59.5	63.8
Golf	8.0	16.8	22.4	29.4
Tennis	9.0	10.7	14.2	24.0
Weight training	5.2	13.4	13.4	17.0
Baseball, softball	23.6	20.2	14.2	14.6
Social dancing	25.7	39.5	39.5	35.9
Fishing	4.5	4.7	4.7	4.5
Volleyball	5.0	4.0	4.5	4.7
Bowling	19.1	20.9	18.7	13.7

Values are percentage participation in a 12-month period.
Source: Canadian Fitness & Lifestyle Research Institute; Leisure Studies Data Bank, University of Waterloo (Table 3.1: 12).

Table 3.12. Participation (%) in selected leisure and cultural activities by education and occupation: Canada, Jan–March, 1972.

	Education			Occupation			
	Elementary	Secondary	University	Unemployed	Blue Collar	White Collar	Professional
Watching TV	92	96	93	95	95	94	93
Listening to records	32	56	62	48	56	63	53
Attending movies	16	46	57	34	44	54	39
Attending non-classical music concert	5	14	16	17	11	16	16
Attending sports events	11	26	28	25	26	28	23
Attending live theatre	4	12	25	14	6	11	19
Attending opera or operetta	1	2	6	1	1	2	4
Attending ballet performance	0	1	3	1	0	1	2
Attending concert of classical music	2	7	18	6	3	7	16
Visiting public art gallery*	2	6	19	6	4	6	15
Visiting museum*	3	8	17	8	4	8	14
Visiting historic site*	3	13	22	10	7	12	17
Visiting fair, exhibition, carnival	7	13	14	11	13	14	10
Engaging in arts, crafts or music	15	28	38	18	26	37	26
Hobbies	5	14	24	9	13	20	14
Participating in sports activities	10	27	33	22	26	32	24
Participating in physical activities	8	23	34	12	22	29	23

Source: Statistics Canada; Leisure Study – Canada 1972 (Table 3.1: 3).
* Sum of free and paid attendance.

Table 3.14, based on time estimates reported in the 1988 Campbell's *Survey of Fitness and Well-being* (Canadian Fitness and Lifestyle Research Institute, 1991), suggests that greater access of the lower SES groups to free time is consumed primarily in such activities as watching TV, hobbies, and possibly social activities, while higher SES groups spend more time reading and engaging in cultural activities. Table 3.13 also shows that free time of higher status groups is, in relative terms, more heavily weighted toward cultural activities, reading, and social and physically active leisure, while lower status groups spend more free time in mass media consumption (watching television), rest, relaxation, and hobbies – the less physically active forms of leisure.

In general, the findings on socioeconomic status differences in the distribution of free time seem to support Wilensky's (1963) observation that Veblen's (1899) notion of the 'leisure classes' does not apply to contemporary North America. In modern societies, higher socioeconomic status usually signifies greater access to financial and information resources, yet at the same time it is associated with functional responsibilities and career advancement expectations which predicate more intensive job involvement accompanied by greater time pressure. To paraphrase Wilensky: with economic growth upper strata have probably lost rather than gained leisure.

TRENDS IN LEISURE PARTICIPATION AND THE USE OF TIME

Dumazedier (1967), Linder (1970), Robinson and Converse (1972), Robinson (1987), Schor (1991) and Gershuny (1992) have provided different answers to the question of whether, in the last two or three decades, the lives of people in advanced industrial societies have become more 'leisurely' or more 'harried', more free or more constrained. It seems that the answer to these questions, similar to that about social differences in leisure, depends on the measurements of leisure used for the analysis of trends.

Trends in Leisure Participation

In general, leisure participation data do not lend themselves readily to the analysis of historical trends and lifestyle changes because of differences in survey design. However, some examination of trends is possible in the Canadian context. A comparison of data from the *Leisure Study – Canada 1972* (Kirsh *et al.*, 1973) and *Leisure Study – Canada 1975* (Schliewen, 1977)

Table 3.13. Allocation of time (minutes per day) to major groups of activities by socioeconomic status: Canada, 1986.

	Paid work	Domestic work*	Sleep	Free time	Free time activities*						
					TV	Social leisure	Hobbies	Rest, relax	Reading	Active leisure	Culture
Blishen's scale											
High SES†	419	83	467	244	79	81	6	18	16	15	11
Medium SES†	414	93	464	262	105	79	11	18	13	14	7
Low SES†	371	99	485	278	121	80	13	23	10	15	6
Pineo–Porter–McRoberts' occupational groupings											
Professionals	412	89	460	259	78	89	11	18	19	15	10
High & middle management	451	69	455	252	100	75	9	18	19	13	6
Clerical	363	112	483	268	107	83	14	20	14	11	6
Blue-collar	404	93	476	272	131	71	11	23	8	15	5

* *Domestic work* includes household chores, shopping and child-care; *social leisure* includes visiting friends, entertaining at home, and outings; *reading* includes reading of books and magazines; *active leisure* includes physically active, outdoor and sporting pursuits; *culture* includes attending performing arts, visiting museums and galleries.

Source: GSS, Statistics Canada; Research Group on Leisure and Cultural Development, University of Waterloo (Table 3.1: 14).
† SES, socioeconomic status.

Jiri Zuzanek

Table 3.14. Estimated time (hours per week) spent in selected leisure activities by gender, age and education, Canada, 1988.

	Total	Gender		Age					Education			
		Men	Women	<19	20–24	25–44	45–66	65+	Less than high school	High school	Some post high	University
Watching TV	9.1	9.4	8.9	9.3	8.0	8.6	9.6	11.2	10.5	9.8	8.6	7.3
Reading	5.9	5.5	6.3	4.5	6.1	5.6	6.4	8.6	4.9	4.9	7.0	8.1
Hobbies and crafts	3.5	3.5	3.5	3.3	3.0	3.1	4.0	5.4	3.4	3.6	3.9	3.1
Visiting with relatives	2.4	3.1	2.8	2.1	3.0	2.8	2.8	4.0	2.9	3.0	2.7	2.3
Visiting with friends	4.9	4.6	4.7	7.7	6.5	3.9	3.4	3.8	5.2	5.0	4.8	3.9
Attending cultural events	0.7	0.8	0.7	0.9	0.9	0.7	0.7	0.6	0.7	0.6	0.8	1.1

Source: Canadian Fitness & Lifestyle Research Institute; Leisure Studies Data Bank. University of Waterloo (Table 3.1: 12).

suggests that, for most leisure activities, participation rates in 1975 were considerably higher than in 1972 (Tables 3.2 and 3.10). Yet this increase, in all likelihood, does not reflect lifestyle or social change, but rather is a product of a different 'time referent' used in the two surveys (see Table 3.1, column 8). In 1972, respondents were asked to report their participation in selected leisure activities for a period of approximately *2 months*, from January to March 1972, while in 1975 participation for a period of *12 months* preceding the survey was recorded. Clearly, the chances of participating in an activity are greater over a full year than in 2 winter months. Comparing Tables 3.2 and 3.10 it seems that in *most* activities the 1975 participation rate is higher than for 1972.

Leisure Study – Canada 1975 and the 1978 *Survey of Canadians' Reading Habits* used the same time referent, recording participation for a period of 12 months, but produced vastly different results with regard to participation in the same activities. In general, levels of participation reported in the 1978 survey were *higher* than in the 1975 survey. According to the *Leisure Study – Canada 1975*, 27% of Canadians reported visiting a library at least once during the year preceding the survey; in 1978, this percentage was 43.5. In 1975, 13% of Canadians attended a live theatre performance, and 8% reported attending classical music performances; in 1978, the respective figures were 30.7% and 23.4%.

On the other hand, according to the 1981 *Fitness Canada Survey*, Canadians' participation in physically active leisure and sporting activities declined substantially during the 5-year period from 1976 to 1981. According to the 1976 *Survey of Fitness, Physical Recreation and Sport*, 33.6% of Canadians reported ice-skating at least once during the year preceding the survey; in 1981, the percentage was 23.5. In 1976, 24.9% of Canadians reported playing tennis; in 1981 the figure was 16.9%. In 1976, 16.7% of respondents reported playing ice-hockey; in 1981 only 10.4% did.

At first glance, we are confronted with a dramatic shift of Canadians' interests and participation from the sports to the arts. Yet these figures are not substantiated by other evidence, such as box office returns, and as suggested below, probably reflect differences in survey design rather than social change.

Higher rates of cultural participation reported in the 1978 *Readership Survey*, may in part result from the sequencing of this survey's general and specific questions being different from that in 1975. In the *Leisure Study – Canada 1975*, questions about leisure participation were asked at the beginning of the survey in a rather straightforward manner: 'Have you participated in any of the following leisure activities in the last 12 months?' A list of activities followed. If a person responded 'yes', he or she was further asked how many hours were spent in each of the reported activities during a typical week in July–August and September–October of 1975. This 'matter of fact' sequencing of questions, with a follow-up question checking on the validity of the original answer, may have implicitly discouraged respondents from exaggerating their participation in surveyed activities.

In 1978, the sequence of general and specific questions was reversed. Respondents were first asked how many hours they spent in a given activity during the last week. For activities where no participation was reported, a question followed: 'Have you participated in this activity in the last 12 months?'. The 1978 questionnaire 'loosened' rather than 'tightened' the grip in the follow-up part of the question and, arguably, gave respondents a leeway to answer the questions in a 'normatively expected' rather than objective manner. Obviously this proposition needs to be put to an empirical test, yet it is consistent with an argument made by Stynes *et al.* (1980), that even small nuances in the formatting of questions often result in considerable inconsistencies in findings.

The lower rates of physically active and sporting participation in the 1981 *Fitness Canada Survey*, compared with the 1976 *Survey of Fitness, Physical Recreation and Sport*, may also have been caused by differences in the design and formatting of the surveys' questionnaires. In the 1976 survey, respondents were given a list of activities and asked to indicate whether they participated in any of the listed activities during the last 12 months. In 1981, respondents had to name the activities in which they participated, *without the help of an activity list*. Recalling past participation without the assistance of a check list is rather burdensome. Conceivably, the failure of an unprompted recall may have contributed to lower participation rates in 1981.

An analysis of leisure participation surveys administered by different agencies, using different research strategies, poses great difficulties. To what should one attribute, for example, a difference of almost 20% in movie attendance as reported by three national leisure participation surveys conducted within a 2-year period between 1990 and 1992 (Tables 3.6, 3.15, 3.16)? Or, how can one explain the difference between the 3% and 11% attendance at ballet performances as reported by the 1992 *GSS* and the 1992 *PMB Consumer Product Preference* surveys? How many Canadians play golf? Is the number 23%, as suggested by the *Canadian Arts Consumer Profile Survey* (Table 3.15), or a mere 7.4%, as reported by the *PMB Survey of Canadian Consumer Product Preference* (Table 3.16)? Unfortunately, there appears to be no explanation for these large discrepancies.

Even surveys designed along similar lines have often failed the test of reliable across-time comparison. Findings from the 1981 *Canada Fitness Survey* and its 1988 follow-up show that participation in most surveyed leisure activities showed a moderate increase during the period separating the two surveys. Yet for some activities the increase was far too large to be credible: for example, the almost three-fold increase in social dancing and the two-fold increase in gardening (Table 3.5).

To provide reliable indicators of trends in leisure participation, it will be necessary to standardize surveys to a considerably greater extent than has been the case in the past. Canadian data on leisure participation available to date are too 'soft' to allow credible trend analyses. Based on existing data, it is hard to tell whether Canadian society, in the last 20 years, became more passive or more active, more social or more 'cocooned', or whether people read less or more. If leisure participation surveys are to be believed, there was no decline in movie attendance over the past 20 years. Yet this conclusion contrasts starkly with the evidence of declining movie theatre ticket sales. In sum, existing leisure participation data do not possess the precision, con-

Table 3.15. Participation in selected leisure activities: Canadian Arts Consumer Profile Survey, 1990–1991.

Activities	Percentage participating in 12-month period
Watching video	87
Listening to records, tapes, CD	88
Reading books	85
Going to a library	56
Going to movies	67
Attending a sporting event	48
Attending concert/performance	67
Visit museum/art gallery	51
Visiting amusement/theme park	48
Camping/hiking	47
Bicycling	43
Fishing/hunting	31
Whale/bird watching	14
Cross-country/downhill skiing	28
Playing racquet sports	18
Golfing	23
Playing musical instrument	22
Singing in a choir	6
Taking part in live performance	24
Photography	34
Gardening (in season)	56
Dining out	96
Drinking at a bar	62
Entertaining at home	86
Visiting friends/relatives	99
Playing bingo	16
Pool, darts, bowling	46
Working as a volunteer	37

Source: Canada Arts Consumer Profile, 1990–1991 (Table 3.1: 13).

Jiri Zuzanek

sistency and 'depth' to form a basis for effective analyses of trends in the domain of leisure.

Table 3.16. Participation in selected leisure activities by gender: PMB Canadian Consumer Product Preferences Survey, 1992.

Activities	Percentage participating in 12-month period		
	Total	Men	Women
Going to movies	58	58.0	57.0
Attending live theatre performance	34	29.9	38.6
Attending ballet performance	11	8.6	14.1
Attending classical music concert	13	10.9	14.6
Attending jazz concert	7	7.0	6.5
Going to horse races	10	10.4	8.9
Going to baseball game	25	30.2	19.8
Going to football game	10	14.7	5.5
Going to hockey game	25	34.0	17.3
Going to soccer game	4	5.1	3.6
Going to tennis match	3	3.8	2.6
Exercise at home	19	27.1	11.9
Jogging	23	27.0	18.4
Swimming	49	47.8	49.2
Bicycling	30	33.7	26.3
Boating	17	20.9	13.6
Sailing	7	8.0	5.4
Cross-country skiing	17	16.7	17.3
Downhill skiing	17	20.7	13.9
Skating	22	22.5	21.8
Golfing	7	9.2	5.8
Playing tennis	12	15.8	8.3
Playing squash	14	19.2	9.4
Playing basketball	22	29.9	13.6
Playing baseball	16	22.4	16.1
Playing ice hockey	10	18.0	2.3
Curling	6	7.6	3.8
Bowling	23	24.0	22.2
Woodworking	21	35.5	7.0
Gardening	52	48.0	54.9
Knitting	18	2.7	32.7
Entertaining at home	68	63.4	72.3
Dancing	48	45.3	49.7
Dining out (high quality)	52	51.0	53.3

Source: PMB Product Profile Survey, 1992 (Table 3.1: 15).

Trends in the Use of Time

Time-budget studies seem to operate on firmer ground than leisure participation surveys when analysing social trends. A comparison of the use of time by comparable life cycle groups in 1981 and 1986 suggests that the amount of free time available to Canada's employed population remained relatively stable or possibly even declined between the years 1981 and 1986 (Tables 3.7, 3.8 and 3.17). Over the period from 1981 to 1986, free time appears to have decreased in Canada. While data directly comparable to the 1986 survey were not available for 1981, a comparison of matching subsamples showed a decrease in free time over the five year period by over $\frac{1}{2}$ hour per day or 4.2 hours per week. The decrease was virtually identical for women and men (Harvey *et al.*, 1991).

Changes between 1986 and 1992 are relatively small, and also do not substantiate the forecasts of an impending 'society of leisure'. Table 3.17 indicates that for the population as a whole during the 6-year period from 1986 to 1992, work for pay increased by less than 2 minutes per day while domestic work increased by 16 minutes per day. The amount of time spent in personal needs declined by 6 minutes per day, while the amount of time spent in free time activities has increased by 4 minutes per day (less than 1%).

Tables 3.8 and 3.17 also indicate that, while the total amount of free time remained approximately the same in 1992 and 1986, noticeable changes may have occurred in its distribution between various components and among different life cycle groups. In 1992, Canadians seemed to have spent more time than in 1986 in social and physically active leisure. The amount of time spent on rest, relaxation, and watching TV, on the other hand, declined slightly during this same period (Table 3.17). Married women, both working and those staying at home, appeared to be the net 'losers' with regard to free time. Working mothers with small children had 11% less free time in 1992 than in 1986. Married working women aged 35 to 64 with no children at home, had almost 10% less free time in 1992 than in 1986. For homemakers aged 25 to 44, with small children at home, free time declined by 10% between 1986 and 1992. On the other hand, single and retired women had slightly higher amounts of free time in 1992 than in 1986. Among men, changes in the amounts of free time available to different life cycle groups were not pronounced, and did not form a clear pattern (Table 3.8).

Perceived changes in time pressure in 1992, as compared to the situation '5 years ago', are summarized in Table 3.9. This table suggests that most Canadians felt less leisurely in the early 1990s than they did in mid-1980s. The table reflects a change in historical as well as life cycle time. Not surprisingly, respondents who moved to a less time constrained life cycle situation, such as retirees, reported a considerable decline of time pressure in 1992, compared with 5 years earlier – by 50% for retired men and 47% for retired women. Yet every other life cycle group reported an *increase* rather than a decline in perceived time pressure. Some 84.6% of employed mothers aged 25–44 with small children, and 67% of single employed women under 35 years of age reported an increase in perceived time pressure, while 62.5% of employed fathers with small children, and 60.6% of single employed men under the age of 35 reported feeling more rushed in 1992 compared with 5 years earlier. While some of this increase could be attributed to life cycle progression, as in the case of employed parents with small children, an increased feeling of being 'always rushed' among single men and women probably should be attributed to an 'across-the-board' acceleration of the perceived rhythm of everyday life.

The findings on changes in the use of time, are more consistent than the findings on trends in leisure participation. However, the period for which time-

Table 3.17. Changes in duration (minutes per day) of selected activities – 1986–1992 – General Social Surveys, Canada.

	1986		1992	
	Mean	s.d.	Mean	s.d.
Work for pay	193	254	195	261
General domestic work/child care	111	135	127	141
Domestic work	111	135	127	141
Shopping and services	49	68	50	62
Primary child care	53	185	26	70
Personal needs	608	148	602	124
Night time sleep	483	136	483	111
Free time	353	221	357	222
Social leisure	70	119	84	125
Popular culture/spectatorship	5	31	5	34
Arts and culture	1	9	1	10
Physically active leisure	20	60	23	63
Hobbies	16	61	14	60
Pastimes	9	42	11	47
Rest and relaxation	30	79	26	64
Television viewing	135	141	131	135
Listening to radio/records	5	31	5	30
Reading newspapers	10	29	12	30
Reading books and magazines	16	52	17	49
Interpersonal communications	17	47	16	39

Source: GSS, Statistics Canada; Research Group on Leisure & Cultural Development Waterloo (Table 3.1: 11, 14).

use comparisons can be made for Canada, is relatively short. Differences in sampling and coding may account for some of the unexpected decline of free time between 1981 and 1986. Yet the main problem facing time budget researchers in examining social trends is perhaps that of interpretation rather than inconsistencies in findings. For example, how is the discrepancy between a seeming increase of free time (more apparent in the US than Canadian surveys), and the growing sense of subjectively perceived time pressure on the part of large segments of North America's population to be explained?

LEISURE PARTICIPATION AND THE USE OF TIME IN A COMPARATIVE PERSPECTIVE: USA AND CANADA

The task of interpreting relationships between the use of leisure and selected demographic or socioeconomic indicators, as well as leisure trends in Canada, could be greatly assisted by comparing Canadian findings with those from other countries. Comparisons with the findings collected over the last 30 years by US researchers appear particularly useful.

US data on participation in outdoor recreational and cultural activities from the late 1970s and mid-1980s suggest that social differences in leisure behaviour in the US follow similar patterns to those observed in Canada. Table 3.18 indicates that, with few exceptions, men in the US participate in outdoor recreational activities more actively than women, while the opposite seems to be the case with most cultural activities (Table 3.19).

Progression through the life cycle is associated, as in Canada, with a relatively steep decline of participation in physically demanding and sporting activities, yet has little effect on participation in such outdoor activities as fishing, driving for pleasure, sightseeing, walking, birdwatching, picnicking, etc. (Table 3.18). Participation in 'high culture' peaks in the middle stages of the life cycle, but tapers off after the age of 55 (Table 3.19). Higher educational and income status correlate positively with participation in most recreational and leisure activities with the possible exception of such activities as fishing and hunting (Table 3.18).

In an attempt to establish historical trends in leisure participation, Robinson (1979,1987) and Caplow *et al.* (1991) compared data on participation in outdoor and sporting activities from US National Recreation Surveys of 1960, 1965, 1972, 1977 and 1982. Similar to Canada, conclusions to be drawn from data in the US surveys appear rather contradictory. According to one set of data, participation in pleasure driving, picnicking, and sightseeing increased between 1960 and 1977, while according to another set of data, it fell between 1965 and 1982 (Table 3.20). It is not easy to account for fluctuations between an 8% participation rate for tennis playing reported in 1972, compared with 33% in 1977, and only 16% in 1982, or a decline in fishing

from 53% in 1977 to 34% in 1982 (Table 3.20). Caplow *et al.* acknowledge
that national series of data on participation in important leisure and cultural
activities are 'sparse and unsatisfactory' and that, while surveys of leisure
participation are useful for cross-sectional analyses, they are 'defective for
delineating trends' (1991: pp. 462–467).

Questions on whether people living in advanced industrial societies have
greater access to free time or live more 'harriedly' have been addressed by a
number of US authors, including Juster (1985), Robinson (1985, 1987,
1990), Cutler (1990) and Caplow *et al.* (1991). The results of some of these

Table 3.18. Participation in selected outdoor recreational activities by gender, age and
education: US, 1977.

	Total	\ Men	Gender\ Women	Age\ 18–24	35–44	55–64	65+	Education*	Income*
				Percentage participating in a 12-month period					
Bicycling	47	46	47	67	46	16	10	–	–
Horseback riding	15	16	14	20	9	2	1	–	–
Golfing	16	22	11	18	15	19	8	0.08	0.18
Tennis	33	33	32	53	34	7	2	0.15	0.1
Canoeing, kayaking	16	18	14	27	14	3	2	–	–
Sailing	11	11	11	17	12	4	2	0.13	0.1
Waterskiing	16	20	12	31	17	3	0	–	–
Swimming, sunbathing (pool)	63	48	44	82	73	41	19	0.09	0.12
Fishing	53	65	43	39	37	46	30	-0.09	–
Hunting	19	33	7	25	27	11	2	-0.14	–
Hiking & backpacking	28	34	23	40	32	14	7	0.15	–
Nature walks, birdwatching, photo	50	47	53	47	62	43	33	–	–
Picnicking	72	71	74	74	81	64	47	0.08	–
Driving for pleasure	69	69	70	83	79	66	51	–	–
Sightseeing	62	63	62	63	73	59	44	0.15	0.11
Attending outdoor sports events	61	67	56	74	67	49	30	–	–
Ice skating (outdoors)	16	17	16	22	18	4	1	–	–
Cross-country skiing	2	3	2	4	3	1	0	–	–
Downhill skiing	7	8	6	14	7	1	0	0.09	–

Source: The *Third Nationwide Outdoor Recreation Plan*, Appendix II, Robinson, 1979, pp. 79–85.
* Value (beta) represents relative contribution of education and income to leisure participation,
after contributions of other variables are taken into account.

Table 3.19. Participation in selected cultural activities by gender, age and education: US, 1985.

	Percentage participating in a 12-month period						
	Plays	Musicals	Opera	Ballet	Classical music	Jazz	Museums
Total	12	18	3	4	13	10	22
Men	11	16	3	3	11	10	21
Women	13	20	3	6	14	9	23
18–24	11	16	2	4	12	16	22
25–34	12	18	2	5	12	15	26
35–44	15	22	4	6	16	9	27
45–54	13	20	4	4	15	8	22
55–64	11	18	3	4	2	5	19
65–74	10	14	3	3	13	3	15
75+	0	9	2	2	8	1	9
Grade school	2	4	1	1	2	1	3
High school	7	13	2	2	7	7	15
University	26	36	7	10	29	19	45
Graduate school	36	42	10	14	40	22	56

Source: Robinson, 1987, p. 45.

analyses are summarized in Tables 3.21 and 3.22. According to Table 3.21, time spent in work for pay, for respondents in the 25–44 age group, did not change significantly in the period from 1965 to 1981. The trends with regard to free time are not very clear. The amount of free time available to Americans seems to have increased substantially, for those aged 25–44, from 1965 to 1975, but changed relatively little thereafter (Tables 3.21 and 3.22). The picture becomes even more complex when the analyses focus on changes in the allocation of time by men and women in the age group 45 to 64. For this age group the amount of leisure time declined between 1975 and 1981 for men, and slightly increased for women.

As for the subjective perception of being 'always rushed', US time-use surveys of 1965, 1975 and 1985 suggest that, in spite of some increases of free time during this period, American men and women felt subjectively more pressed for time in 1985 than in 1965 (Robinson, 1990). The proportion feeling 'always rushed' increased for men from 23% in 1965 to 27% in 1975, and to 30% in 1985. For women the increase was from 26% in 1965 to 28% in 1975, and 33% in 1985 (Table 3.22).

In general, US time-budget findings parallel those of Canada, and provide additional evidence of the complexity of life in modern industrial societies marked, as it seems, by concurrent trends toward greater time freedom as well as 'harriedness' and 'time pressure'.

CONCEPTUAL, POLITICAL, AND METHODOLOGICAL ISSUES OF TIME-USE AND LEISURE RESEARCH

Canada has a respectable record of leisure participation and time-budget surveys. But, while the many surveys conducted over the last 25 years have answered a number of questions, they also leave many questions unanswered. The questions faced by researchers in other countries also remain unresolved in Canada. These questions can be considered as conceptual, practical, political, methodological, operational and substantive.

Table 3.20. Participation in selected outdoor recreational activities: US, 1965, 1970, 1972, 1977 and 1982.

	Percentage of participation for a 12-month period				
	1965	1970	1972	1977	1982
Bicycling	19	19	20	47	32
Horseback riding	10	9	8	15	9
Golfing§	9	–	6	16	10
Tennis§	6	–	8	33	16
Boating	26	24	17	34	22
Canoeing	4	–	4	16	8
Sailing	3	–	3	11	6
Waterskiing	7	–	5	16	9
Swimming*	49	–	21	63	53
Fishing	33	28	30	53	34
Hunting†	13	–	14	19	12
Camping‡	12	20	13	30	23
Hiking	9	–	7	28	15
Walking for pleasure	51	30	44	68	53
Birdwatching	6	4	7	–	12
Picnicking	60	48	51	72	48
Pleasure driving	59	–	49	69	48
Sightseeing	54	–	44	62	46
Attending outdoor sports	42	35	23	61	40
Ice skating†	9	–	–	–	6
Skiing†	4	–	–	–	9
Sledding†	13	–	–	–	10

Sources: Third Nationwide Outdoor Recreation Plan, Appendix II, 1979, p. 111; Robinson, 1987, p. 36.
* Data for 1972 and 1977 refer to outdoor pool swimming.
† For nine non-summer months.
‡ Data for 1972 and 1977 refer to camping on developed camp grounds.
§ Summer months only.

Conceptual Issues

Conceptually, one of the main issues arising in the conduct of both leisure participation and time-use surveys is how to develop a functionally relevant classification of daily and leisure activities that can form a valid base for large scale national leisure and time-use surveys. In the publication *Human Activity Patterns in the City*, Chapin (1974) suggested that an activity class might be, for example, shopping. However, shopping can also be defined as: driving from home to shopping centre, buying groceries, and driving back home. The same activity may be classified in even greater detail, for example: (i) driving from home to the shopping centre; (ii) hunting for a parking space; (iii) picking up a cart, and so on. The problem of leisure and time-use research is how to define an 'activity universe' that is both comprehensive and manageable. How many activities should be on the check list? How detailed should the activity codes

Table 3.21. Allocation of time to major groups of activities, persons aged 25–44: US, 1965–1981.

	Hours per week			
	1965[1]	1975[1]	1975[2]	1981[2]
Men				
Work for pay	41.8	41.2	41.5	40.9
Travel to work	4.4	4.6	3.8	3.6
Household work	11.8	8.9	11.2	14.0
Personal care (sleep and eating)	67.3	65.5	68.5	65.9
Free time	38.7	44.5	40.0	40.6
Television	12.3	14.5	12.6	12.1
Reading	4.4	3.6	2.8	2.7
Social entertainment	12.8	14.8	12.8	12.4
Active leisure	3.0	5.1	4.6	5.4
Other	4.1	3.3	3.1	2.9
Women				
Work for pay	13.5	16.5	17.8	21.0
Travel to work	1.3	1.4	1.5	2.2
Household work	45.9	35.4	33.2	31.8
Personal care (sleep and eating)	70.9	72.4	72.0	70.5
Free time	35.2	41.0	41.9	40.5
Television	8.9	14.3	12.6	11.0
Reading	3.4	2.8	3.2	3.0
Social entertainment	11.7	11.1	13.9	14.4
Active leisure	2.8	3.5	4.0	4.2
Other	1.2	1.3	1.6	1.9

Source: Juster, 1985, pp. 316–317; 320–321.
[1] Urban households only.
[2] Urban and rural households.

Table 3.22. Time spent in leisure activities, persons aged 18–64: US, 1965, 1975 and 1985.

	Hours per week		
	1965	1975	1985
Total free time	34.5	38.3	40.1
Television	10.5	15.2	15.1
Reading	3.7	3.1	2.8
Visiting, talking	9.2	7.7	9.2
Sports, outdoors	0.9	1.5	2.2
Hobbies	2.1	2.3	2.2
Thinking, relaxing	0.5	1.1	1.0
Cultural events	0.8	0.5	1.1
Clubs and organizations	1.0	1.2	0.7
	% Feeling 'always rushed'		
Men	23	27	30
Women	26	28	33

Source: Cutler, American Demographics, Nov. 1990; Robinson, *American Demographics*, 1990.

be? How significant is the difference between camping at developed sites as opposed to the undeveloped ones, or swimming in outdoor pools as opposed to 'other' outdoor swimming? Should we combine baseball and softball into one activity category or treat them separately? Should attendance at popular and rock concerts be one activity or two? What about billiards and pool? How long can the list of recreational activities be, before it alienates the respondent? Without resolving some of these issues, practical relevancy and methodological consistency will remain hard to achieve.

Practical Issues

Practically, time-use surveys provide effective information about the distribution of time among the most common, frequently engaged in, daily and leisure activities, but are of limited use for monitoring participation in 'lifestyle specific' leisure pursuits. Time-use surveys can be relied upon for information on such common activities as watching television, reading for pleasure, social leisure (visiting friends and entertaining at home), yet offer rather patchy information on arts attendance or sports participation because they are not necessarily 'everyday' activities. It is also noticeable that, with the proliferation of telephone interviewing, the mean number of activities reported by time-budget respondents has declined over the years. As a result, the emerg-

ing picture of daily behaviour becomes increasingly 'broad stroke' and approx-
imate, occasionally blurring significant differences in human behaviour.

Leisure participation surveys, by contrast, are well suited for the analysis
of less frequent but 'focused' activities, such as active sports or culture. Yet
they are almost entirely useless in examining participation in the more
'routine' leisure activities in any meaningful way. For example, what is there
to be gained from knowing that approximately 95% of all respondents in
Canada and the US, men and women, young and old, watched television at
least once during the past 12 months?

Political Issues

Politically, leisure participation surveys can be described as 'inflationary',
while time-use surveys can be seen as 'deflationary' or 'conservative'. When
responding to a checklist of leisure and cultural activities, respondents often
succumb unconsciously to the normative expectations of society, and over-
report participation in more desirable, socially approved or prestigious activities. 'Is it possible that I haven't been to the theatre this year? Surely, I must
have been at least once, but have simply forgotten. Let's check it off. It will
please the research team, anyhow, won't it?' (Since surveys rarely delve into
socially less 'approved' activities, such as gambling and drinking, the opposite
tendency – to under-report activities – is less apparent.) On the other hand, the
very design of time-use diaries, with their focus on tedious reconstruction of
the sequence of daily events, seems to take normative pressures off the
respondent's shoulders and the shorter, specific time period involved gives less
scope for inaccurate recall of events that 'might have been'. Not surprisingly,
time-budget data show greater across-survey and across-time consistency; yet
by apparently under-playing participation in worthy leisure and cultural
pursuits, they provide little political 'ammunition' to leisure and culture
lobbyists.

An example of this phenomenon can be given from the 1984 *Kitchener–
Waterloo–Brantford Survey*. Some 46% of respondents claimed to attend at
least one live theatre performance in the previous year, which sounds impressive. However, analyses of 1981 time-budget data show that middle aged
respondents spent an average of only 2 to 5 minutes per day on all 'high
culture' activities (theatre, concerts, opera, art galleries) (Zuzanek and Stewart, 1983). In fact, the latter amounts to between 12 and 30 hours a year,
which is easily enough to accommodate at least one 'high culture' visit – but
the 46% sounds more impressive.

Methodology

Methodologically, the most serious problem facing researchers engaged in cross-sectional and across-time comparisons of leisure participation and use of time is the question of cross-survey consistency. Much confusion in comparing leisure participation and time-use data from different surveys can be attributed to an inconsistent use of *age thresholds* in these surveys. Changing the 'start up' threshold from 14 to 18 years, as a rule, increases overall rates of physically active leisure participation and overall amounts of free time for the total sample, because the 14–18-year-old group are highly active in physical recreation and have more leisure time than the average adult. Thus, in the *Leisure Study – Canada 1975*, attending movies of any type was reported by 63.4% of those aged 14 years or more, but by only 55.3%, if participation were measured for those aged 18 years and more.

Another problem arises from the inconsistency in *defining surveyed activities*. Even surveys focusing upon specific areas of leisure involvement, for example physically active leisure or outdoor participation, are rarely consistent in their choice and definition of surveyed activities. In total, the *CORD* and *TORPS* surveys of the 1970s sought information about participation in 52 outdoor activities, but only seven of these activities were defined in a truly identical fashion. Of 35 cultural activities surveyed in the 1972, 1975 and 1978 Statistics Canada surveys of leisure and cultural participation, only ten were defined identically. Of 39 sporting and fitness activities examined in the 1976 and 1981 surveys of fitness and sports participation, only 13 were worded identically.

For example, while in the *Leisure Study – Canada 1972* respondents were asked about listening to records, tapes and cassettes 'as a conscious leisure activity', the *Leisure Study – Canada 1975* asked respondents about listening to records, tapes and cassettes without any further qualification. Could this have contributed to an apparent 14% increase in record listening in 1975? Is 'visiting an historic site (building, park, etc.)' from the 1972 survey, the same thing as 'visiting a historic site or restoration (pioneer homes and villages, architectural monuments, etc.)' in 1975? Could a seemingly innocuous difference in the wording of the question result in more than a twofold increase in participation, from 10% to 24%, in a period of 3 years? Is this discrepancy a result of seasonal differences, i.e. January to March in 1972, as opposed to the whole year in 1975? Or is the change due to the difference in the length of 'time referent', i.e. 3 months in 1972 as opposed to 12 months in 1975 (Tables 3.2 and 3.10). Similar observations can be made for US surveys, where frequent changes in the wording of activity related questions have made trend analyses in many instances all but impossible. According to Stynes, Bevins and Brown, 'When definitions or groupings of activities change even slightly from survey to survey or year to year, few clues about recreation trends for these activities can be discovered' (1980, p. 226).

For time-budget surveys, one of the most important methodological problems is the question of consistent *coding* and grouping of activities. Frequent changes in the definition, coding and grouping of activities are among the most common reasons for discrepancies in survey findings. While changes in grouping and classifying of activities can be rectified by appropriate recodes, different coding instructions may irreparably reduce across-survey comparability. It is likely that lower figures for child care and higher figures for social leisure in the 1992 GSS time-use survey, compared with 1986, are due primarily to differences in coding.

Operational Issues

Operationally, surveys which focus on selected areas of leisure or recreational participation, such as physically active leisure or arts, often produce a 'magnifying glass effect' – that is, they encourage over-reporting of surveyed activities. This may have contributed to an apparent increase in public library attendance from 27% in 1972 to 40% in 1978, since Statistics Canada's 1978 survey focused specifically on *reading habits* of Canadians. This factor may also explain lower rates of sports participation in the 1992 *PMB Consumer Preferences Survey* (Table 3.16), compared with the Canada Fitness surveys of the 1980s (Tables 3.5, 3.11 and 3.16)

Organization

Organizationally, two major challenges facing those responsible for large scale leisure participation and time-use surveys are, first, continuity of research effort and second, the establishment of a clearly defined research focus as opposed to an 'all-purpose' or 'omnibus' design. It has been suggested that two factors which have negatively affected the study of leisure and culture in Canada can be termed the 'novelty' and 'gypsy' syndromes (Zuzanek, 1988). The virtues expected of large scale statistical surveys are not necessarily novelty and experimentation, but rather across-survey and across-time consistency. Yet within Canada's federal organizations frequent changes of mandates ('gypsy' syndrome) and ambitions to put 'a new mark' on earlier efforts ('novelty' syndrome), have often resulted in unwarranted design changes. This was particularly true of leisure and cultural participation studies, and to a lesser extent of time-use research.

Another problem which has plagued many large scale leisure and cultural participation surveys has been their attempt to serve too many diffuse purposes and constituencies. Government departments and statistical agencies have often generated massive amounts of data of little use to any particular constituency. Post-factum efforts to entice researchers to use exist-ing data for secondary analyses have on some occasions provided a much needed research focus.

Substantive Issues

Substantively, time-use surveys provide information about the distribution of free time as a leisure resource and *opportunity*. However, they do not tell us how actively this opportunity is used, nor how the available free time is structured. This information, at least in part, is provided by leisure participa-tion surveys. From this perspective, data on leisure and recreational participa-tion serve as an indication of respondents' and groups of respondents' *interest* in, and *competence* to live up to the challenge of free time. Time-use and leisure participation surveys provide *complementary* rather than competing strategies for studying leisure behaviour.

IN CONCLUSION

Leisure research, whether in the form of leisure participation or time-use studies, can benefit greatly from two strategies: learning from past experience and mistakes, and making greater use of small scale experimental designs to test the validity of and develop optimal designs for research instruments.

Statistical agencies should pay greater attention to what was done in the past: what was successful and what was not successful. One of the most striking observations arising from analyses of past leisure studies is how remarkably short the 'corporate' memory of major cultural and statistical research agencies is. An important contribution of academic research to leisure and cultural studies may lie in reconstructing and making sense of past effort (rekindling 'corporate research memory'), as well as developing and pretesting, under controlled conditions, better instruments which can benefit large scale leisure studies in the future. 'A clearer picture of the potential influence of methods upon results is best obtained through controlled experi-mental designs' (Stynes *et al.*, 1980, p. 225). A sense that we need to fight conceptual and methodological 'amnesia' in leisure and cultural research and experiment where it is appropriate, i.e. in the focused environment of acade-mia, has inspired these analyses. If this point is noted by other researchers and by statistical agencies we will consider our task worth the effort.

REFERENCES

Blishen, B.R., Carroll, W.K. and Moore, C. (1987) The 1981 Socioeconomic Index for Occupations in Canada. *The Canadian Review of Sociology and Anthropology* 24, 465–488.

Canadian Fitness and Lifestyle Research Institute (1991) *The Well-Being of Canadians. Highlights of the 1988 Campbell's Survey*. Canadian Fitness and Lifestyle Research Institute, Ottawa.

Caplow, T., Bahr, H.M., Modell, J., and Chadwick, B.A. (1991) *Recent Social Trends in the United States: 1960–1990*. Campus Verlag, Frankfurt am Main, and McGill-Queen's University Press, Montreal.

Chapin, F.S., Jr (1974) *Human Activity Patterns in the City*. John Wiley, New York.

Cheek, N. H., Jr and Burch, W.R., Jr (1976) *The Social Organization of Leisure in Human Society*. Harper & Row, New York.

Cutler, B. (1990) Where does the free time go? *American Demographics* 12(11), 24–25.

Decima Research and Les Consultants Cultur Inc. (1992) *Canadian Arts Consumer Profile: 1990–1991. Findings*. Decima Research and Les Consultants Cultur Inc., Ottawa.

Dumazedier, J. (1967) *Toward a Society of Leisure*. The Free Press, New York.

Elliot, D.H., Harvey, A.S. and Prokos, D. (1973) *An Overview of the Halifax Time-Budget Study*. Dalhousie University, Institute of Public Affairs, Halifax, Nova Scotia.

Ennis, P.H. (1968) The definition and measurement of leisure. In: Sheldon, E.B. and Moore, W.E. (eds) *Indicators of Social Change: Concepts and Measurement*. Russell Sage, New York.

Fitness and Amateur Sport (1983) *Fitness and Lifestyle in Canada*. Fitness and Amateur Sport, Ottawa.

Gershuny, J. (1992) Are we running out of time? *Futures* 24, 3–22.

Harvey, A.S. (1991) Spending time. In: McCullough, E.J. and Calder, R.L. (eds) *Time as a Human Resource*. University of Calgary Press, Calgary, pp. 227–242.

Harvey, A.S., Marshall, K. and Frederick, J.A. (1991) *Where Does Time Go?* Statistics Canada, Ottawa.

Juster, T.F. (1985) A note on recent changes in time use. In: Juster, T.F. and Stafford, F.P. (eds) *Time, Goods, and Well-Being*. University of Michigan, Institute for Social Research, Ann Arbor, pp. 313–332.

Kinsley, B.L. and Casserly, C.M. (eds) (1984) *Explorations in Time Use*. Employment and Immigration Canada, Ottawa.

Kirsh, C., Dixon, B., and Bond, M. (1973) *A Leisure Study – Canada 1972*. Culturcan Publications, for the Arts and Culture Branch, Department of the Secretary of State, Ottawa.

Linder, S.B. (1970) *The Harried Leisure Class*. Columbia University Press, New York.

Meissner, M., Humphreys, E.W., Meiss, S.M. and Scheu, W.J. (1975) No exit for wives: sexual division of labour and the cumulation of household demands. *Canadian Review of Sociology and Anthropology* 12, 424–439.

Parks Canada (1976) *Canadian Outdoor Recreation Demand Study. An Overview and Assessment*. Ontario Research Council on Leisure, Toronto.

Pineo, P.C. (1985) *Revisions of the Pineo-Porter-McRoberts Socioeconomic Classification of Occupations for the 1981 Census.* QSEP Research Report # 125, McMaster University, Hamilton.

Robinson, J.P. (1979) The Nationwide Outdoor Recreation Survey: some results and conclusions. In: *The Third Nationwide Outdoor Recreation Plan. Appendix II.* Department of the Interior, Heritage Conservation and Recreation Service, Washington DC, pp. 79–85.

Robinson, J.P. (1985) Changes in time use: an historical overview. In: Juster, T.F. and Stafford, E.P. (eds) *Time, Goods and Well-Being.* University of Michigan, Institute for Social Research, pp. 289–312.

Robinson, J.P. (1987) Time's up. *American Demographics* 9(3), 34–37.

Robinson, J.P. (1990) The time squeeze. *American Demographics* 12(2), 30–33.

Robinson, J.P. and Converse, P. (1972) Social change as reflected in the use of time. In: Campbell, A. and Converse, P. (eds) *The Human Meaning of Social Change.* Russell Sage Foundation, New York, pp. 17–86.

Schliewen, R.E. (1977) *A Leisure Study – Canada 1975.* Comstat Services, Ottawa.

Schor, J. (1991) *The Overworked American.* Basic Books, New York.

Shaw, S.M. (1985) Gender and leisure: inequality in the distribution of leisure time. *Journal of Leisure Research* 17 (4), 266–282.

Statistics Canada (1990) *Time Use, Social Mobility and Language Use – 1986: Public Use Micro Data File Documentation and User's Guide.* Statistics Canada, Ottawa.

Stynes, D.J., Bevins, M.T. and Brown, T.L. (1980) Trends or methodological differences? Paper presented to the *National Outdoor Recreation Trends Symposium,* Durham, NH.

Szalai, A. *et al.* (1972) *The Use of Time.* Mouton, The Hague.

TORPS Committee (1978) Tourism and recreation behaviour of Ontario residents: free time. In: *Tourism and Outdoor Recreation Planning Study. Vol. 4.* TORPS Committee, Queen's Park, Toronto.

Veblen, T. (1899) *The Theory of the Leisure Class.* MacMillan, New York.

Wilensky, H. (1963) The uneven distribution of leisure: the impact of economic growth on 'free time'. In: Smigel, E.O. (ed.) *Work and Leisure.* College and University Press, New Haven, pp. 107–145.

Wippler, R. (1970) Leisure behaviour: a multivariate approach. *Sociologia Neerlandica* 6(1), 51–65.

Zuzanek, J. (1978) Social differences in leisure behaviour: measurement and interpretation. *Leisure Sciences* 1(3), 271–293.

Zuzanek, J. (1979) Leisure and cultural participation as a function of life-cycle. Paper presented to the annual meeting of the *Canadian Sociology and Anthropology Association,* Saskatoon.

Zuzanek, J. (1988) The fads and foibles of cultural statistics. *Journal of Cultural Economics* 14(1), 2–4.

Zuzanek, J. and Box, S.J. (1988) Life course and the daily lives of older adults in Canada. In: Altergott, K. (ed.) *Daily Life in Later Life. Comparative Perspectives.* Sage, Newbury Park, CA.

Zuzanek, J. and Smale, B.J.H. (1992) Life-cycle variations in across-the-week allocation of time to selected daily activities. *Society and Leisure* 15(2), 559–586.

Zuzanek, J. and Stewart, T. (1983) *Mass Media and Arts Participation: Quantitative Dimensions and Life-cycle Variations.* Report to the Department of Communications, Ottawa.

4 France

NICOLE SAMUEL

INTRODUCTION

It was not until after the Second World War that leisure research – including national leisure participation surveys – emerged in France. This became possible for at least two reasons. First, the social sciences, having gained increased recognition and official support, became attractive to a wider number of researchers, whose interest spread from the traditional fields of investigation, such as work, law, religion, the family and education, to new areas of study. Second, leisure had slowly but surely become a social right and a significant social phenomenon in French society (Samuel and Romer, 1984). As a result, knowing about leisure as a social phenomenon *per se* and about leisure participation as a feature of contemporary society, became an important matter, not only for social scientists, but also for planners and politicians and for business people interested in the leisure market. The convergence of these two factors explains why leisure studies came into existence and gathered momentum in the late 1940s, both theoretically and empirically. This take-off was stimulated by the fact that it became possible to obtain statistical information about leisure from research conducted by several institutions, most of which were established during the post-war period.

This chapter presents the contribution of such surveys to the knowledge of leisure participation in France. It includes three sections: first, a brief history and description of French leisure participation surveys; second, a summary of results from both time-budget and participation surveys; and third, some comments about leisure participation today as related to gender, age, size of community, occupation and educational level. Concluding remarks discuss theoretical and methodological problems concerning the conduct of national surveys and their use.

BRIEF HISTORY AND DESCRIPTION OF FRENCH NATIONAL SURVEYS ON LEISURE PARTICIPATION

In this section an overview is presented of the surveys conducted by five different organizations. Basic data on the surveys are summarized in Table 4.1.

The IFOP Surveys

The *Institut Français d'Opinion Publique* (IFOP) was established in 1938, but its first leisure survey was not conducted until 1947. IFOP was a commercial enterprise conducting opinion surveys, founded in Paris by Jean Stoetzel (1910–1987), a social scientist who studied at Columbia University, where he had been most impressed with Gallup's work on public opinion surveys. A journal entitled *Sondages* (surveys), published by the Institute, first appeared in June 1939, just before the outbreak of the Second World War. After the end of the war, two issues of the journal focused on leisure, in 1947 and 1949 (*Sondages*, 1947, 1949). IFOP also carried out vacation surveys in 1961, 1962 and 1963. The first was concerned with the workers in the automobile industry around Paris and in the textile industry in the North of France (*Sondages*, 1962, p. 2). The second, which remained unpublished, dealt with a representative sample of the population in communities of at least 5000 inhabitants and studied their vacations in 1962. The third of these surveys was part of a larger scheme including a survey on preferences for various types of holidays in the event of a reduction of working time, and also a survey on second homes and took place in 1963 (*Sondages*, 1964, p. 2). An effort was made to design the questionnaire so that the answers were comparable with the 1962 IFOP study and with the 1961 INSEE study, described below.

The INSEE Surveys

The *Institut National de Statistique et d'Études Economiques* (INSEE) is one of the public agencies established after the end of the Second World War. Founded in 1946, its aim is to analyse statistics concerning many topics, on a national scale. From INSEE, two types of data gathered through national representative surveys provide information about leisure participation, namely *time-budget*

Table 4.1. Time-budget studies and leisure participation surveys in France.

Survey/Organization	Year	Type	Sample size	Area	Age range	Reports
IFOP – Vacation survey	1961	Quest.	709/514	Paris/North of France	18+	Sondages, 1962, 2
	1962	Quest.	2807	France, urban, 321 sites	18+	
	1963	Quest.	3476	France, 280 sites	18+	Sondages, 1964, 2
INSEE – International time-budget	1965–66	Time-budget	2800	6 towns	18–65	Szalai et al., 1972; Goguel, 1966
INSEE – Time-budget	1967	Time-budget	1795/1073	Paris/Nîmes	18–65	Lemel, 1972; 1974
INSEE – Time-budget	1974–75	Time-budget	6600	France, urban	15+	Huet et al., 1978
INSEE – Time-budget	1985–86	Time-budget	24,400	France, urban + rural	15+	Documentation Française, 1988
INSEE – Leisure behaviour of the French	1967	Quest.	6637	France	14+	Leroux, 1970a, 1970b; Debreu, 1973
INSEE – Leisure behaviour of the French	1987– 88	Quest.	10,872	France	14+	Dumontier et Valdelièvre 1989
INSEE – Vacation surveys	1949	Quest.	6500	18 large cities	14+	–
	1950	"	hholds	"	18+	–
	1951	"	13,200 resp.	"		–
	1957	"	5275	Cities 50k+	14+	Gounot, 1958
	1961	"	13,300	France	14+	Gounot, 1961
	1964	"	17,300	France		Goguel, 1965
	1965–93 (Annual)	"	8000 hholds	France	Adults and children	Mémento du Tourisme, 1993
SOFRES – Trips by French Tourists	(Annual) since 1990	Quest.	10,000	France	15+	SOFRES, Annual
Ministère de la Culture: cultural activities of the French	1973	Quest.	1987	France	15+	Min./Culture, 1974
	1981–82	Quest.	4000	France	15+	Min./Culture, 1982
	1988–89	Quest.	5000	France	15+	Min./Culture, 1990; Donnat and Cogneau, 1990
CREDOC – Living conditions of the French population	1978–81	Quest.	2000	France	18+	CREDOC, 1983
Ministère de la Jeunesse et des Sports – Sports participation survey	1985	Quest.	3000	France	12– 74	Erlinger et al., 1985

CREDOC = Centre de recherche et de documentation sur la consommation; hholds, household; IFOP = Institut Français d'Opinion Publique; INSEE = Institut National de Statistique et d'Études Economiques; Quest., questionnaire; resp., respondents; SOFRES = Société Française d'Études et de Sondages.

studies, which, although obviously aiming at a wider field of investigation, nevertheless provide information on leisure participation in terms of time allocation; and *leisure participation surveys*, which include surveys dealing with leisure activities in general and specifically with vacations.

In France, the preoccupation with time budgets may be traced back to Le Play's research in the nineteenth century (Le Play, 1877). Later this type of study developed in the USSR and in the USA. In France time-budget studies were conducted in the 1940s and the 1950s (Stoetzel, 1948; Fourastié, 1951; Girard, 1958; Girard and Bastide, 1959). But the first such study carried out on a national scale involved participation in the extensive UNESCO funded 'International Time-budget Study of Human Behaviour' conducted in 12 countries on a population of urban and suburban dwellers aged 18–65 (Szalai *et al.*, 1972). The concept of free time was preferred to the concept of leisure. It was defined as the 24 hours of the daily time budget minus work time – professional and domestic – and physiological time (sleep, meals, washing, etc.). This definition, which was adhered to in later French studies, includes within free time, time spent on political and religious activities, as well as time spent on leisure activities.

The French part of this international study was first carried out in six towns, with a total sample of 2800 persons (Goguel, 1966). In 1967, the study was replicated in two urban areas, Paris and Nimes (Lemel, 1972, 1974). In 1974–75 INSEE carried out a new time-budget study in urban areas only, and in 1985–86 it conducted a large scale time-budget study covering the entire country, including the rural areas, and involving a sample of 16,000 households and 24,000 respondents (Documentation Française, 1988). Data were collected in two ways. For frequent activities, one or two persons in each household kept a detailed diary of the use of their time during a given day. For irregular or rather infrequent activities, average frequency of participation over a long period was determined by interview (Grimler and Roy, 1987). In some cases, two activities were reported to have been carried out at the same time: both activities were recorded, one was deemed to be the main one and the other was said to be secondary. The results quoted later only refer to what were recorded as 'main' activities.

INSEE carried out leisure participation surveys in 1967 and in 1987–88. It has also conducted specific surveys about vacations in 1949, 1950, 1951, 1957, 1961 and every year since 1964. The 1967 survey on the *Leisure Behaviour of the French People* took place during the last 3 months of the year with a random sample of 6637 persons. In addition to results about the whole sample, the report presented a breakdown of leisure behaviour according to sociocultural variables such as age, sex, occupation, income, level of education and size of the town of residence. A pre-established list of leisure activities was presented to the respondents, who were requested to indicate whether they were or had been practising each activity, and how often, during the preceding year (Leroux, 1970a and b; Debreu, 1973).

The second INSEE study (1987–1988) took place during a whole year, which made it possible to study the impact of seasons upon leisure activities. It was held on a random sample drawn from 15,000 households (10,872 respondents). The questionnaire was more detailed than the one devised in 1967[1] and included new leisure activities which had appeared in the meantime. Some of the questions were designed to obtain information about the leisure lifestyles of some specific categories of the population (the young, the elderly people). One of the aims of the second study was to attempt a measure of the evolution of leisure behaviour since the first one, 20 years before (Dumontier and Valdelievre, 1989). In both cases, results were tabulated according to sociocultural variables such as gender, age, educational level, occupation and size of the place of residence.

Combining information from three of its own surveys, in 1988 INSEE published a report on the evolution of sporting activities from 1967 to 1984 (Garrigues, 1988).

The INSEE surveys on *vacations* are based on the following definition of vacation:

> ... any trip away from home for at least 4 consecutive days is said to be a
> 'vacation' if it is not taken for reasons of work, study, health care in specialised
> places, illness or death in the family.
>
> (Goguel, 1965)

The INSEE conducted vacation surveys in 1949, 1950 and 1951 in 18 large cities. In 1957, the field of investigation was widened to all cities of more than 50,000 inhabitants – about one-third of the French population at the time (Gounot, 1958). Finally, in October 1961, the INSEE vacation survey included the whole of metropolitan France (rural as well as urban areas), and a sample of 13,300 persons (Gounot, 1961). A similar survey followed in 1964, with a random sample of over 17,000 persons (Goguel, 1965). In addition, since 1965, an annual survey on 'Holidays taken by French Residents' has been carried out in the framework of an INSEE general survey on the economic situation; there are two parts, for winter and summer holidays and the latest available results are for 1993 (Mémento du Tourisme, 1993).

In the period 1990–1993, an organization called the Société Française d'Études et de Sondages (SOFRES), established in 1963, carried out a survey entitled *Trips made by French tourists*. It involved 10,000 persons of French nationality, aged 15 and over, and was performed on a monthly basis, asking the respondents about trips made in France and abroad during the month, whatever the reason for the journey. Thus it did not concern leisure trips only.

The Ministry of Culture Surveys

The Ministry of Culture was founded in 1959. Until 1969, it was headed by André Malraux (1901–1976), a famous writer who later became a Minister in de Gaulle's governments. The Research Department of this Ministry carried out surveys of cultural participation, including some questions on holiday-taking. The first was conducted in October–November 1973, involving a quota sample of some 2000 persons (Ministère de la Culture, 1974). This study was replicated in 1981–82 with a larger sample (Ministère de la Culture, 1982) and again in 1988–89 (Ministère de la Culture, 1990; Donnat and Cogneau, 1990).

The surveys were conducted on the basis of questionnaires administered through interviews at the homes of the respondents. They were based on a wide definition of culture, including participation in sporting activities, do-it-yourself activities, and social leisure activities such as going to the café or belonging to voluntary associations. This was clearly meant to go beyond the traditional French identification of culture with the arts.

The results of these three studies can be compared since the same basic questions were asked to comparable representative samples of persons aged 15 and over. In each case, the results are tabulated for the whole sample according to gender, age, occupation, educational level, marital status and size of the place of residence.

The CREDOC Surveys

The *Centre de Recherche et de Documentation sur la Consommation* (CREDOC) is a public agency, established in 1953. It has a wide field of investigation – the general living conditions of the French population – but its surveys include information on leisure behaviour, including sporting activities and holiday-taking. It publishes a journal called *Consommation*. Upon the request of several bodies (Caisse Nationale des Allocations Familiales, Commissariat Général au Plan, Ministère de l'Environnement, Centre d'Étude des Revenus et des Coûts), in 1978 it started its series of surveys on the living conditions of the French population. In 1981–82 new participants joined the project – among them the recently created Ministry of Free Time, the Directorate of Tourism and the Service for the Study and Planning of Mountain Tourism (SEATM). This explains why more interest in leisure activities and in holiday-taking appeared in the CREDOC surveys from this date. The CREDOC sample (2000 people), drawn through the quota method, is deemed representative of the French population aged 18 and over.

The Ministry of Youth and Sports Surveys

The Ministry of Youth and Sports carried out a survey of participation in sporting activities in 1985. The investigation was conducted by members of the Research Department of the National Institute for Sport and Physical Education (INSEP), an institute that is part of the Ministry of Youth and Sports network. The sample included 3000 persons aged 12 to 74 and a wide definition of sport was adopted, including walking, relaxation exercises, aerobics, gym at home and bathing without necessarily swimming (Erlinger *et al.*, 1985).

RESULTS OF LEISURE PARTICIPATION SURVEYS

Some of the results of the above surveys are presented below, in two sections, relating to time-budget surveys and participation surveys respectively.

Time-budget Studies

As mentioned above, some time budgets were established in France during the 1940s and the 1950s. According to IFOP, leisure time amounted to an average of 2 hours per evening between 1945 and 1948 (Raymond, 1968). From the second IFOP survey we know what were the most frequent occupations during an evening of January 1948. The answers to the question 'What did you do last night?', are as shown in Table 4.2.

Table 4.2. Evening leisure activities, France, January 1948.

	% participating (multiple answers possible)
Leisure	
Reading	24
Radio listening	20
Resting	12
Entertaining/visiting friends	7
Going to a show	6
Playing card games	5
Going to dance hall	1
Other leisure	5
Work	
Professional work	7
Domestic work	23

Source: Raymond, 1968

In 1951, Jean Fourastié gave an estimate of an average of 3 hours of leisure per day for the French adult worker (Fourastié, 1951). On the basis of a non-random sample (120 working families in the Paris area), Chombart de Lauwe's estimation (1958) for the 1950s was from $1\frac{1}{2}$ to 2 hours of daily leisure time for the Paris area; his computation excluded do-it-yourself activities, which he evaluated at an additional average daily $1\frac{1}{2}$ to 2 hours, but he admitted that only part of these activities was perhaps felt to be true leisure. In contrast, Dumazedier (1962) included do-it-yourself activities in leisure time, for which he gave an estimate of between 2 and 3 hours per day in the 1950s, insisting that the leisure time experienced after the midday meal as well as before the evening meal should be added to the evening computation of leisure time suggested by Stoetzel (1948). With regard to free time, the main results of the more systematic INSEE time-budget studies, carried out every 10 years since 1966, are as follows:

- In 1966, the average daily free time for a working man, including adult education and participation in political and religious activities, amounted to 4.1 hours, and for a working woman to 3.7 hours; the average for the whole adult population was 3.9 hours (Goguel, 1966, p. 179).
- In 1974–75, the figures for men and women respectively were 3.2 hours and 2.4 hours (Grimler and Roy, 1987).
- In 1985–86 the respective figures were 3.7 hours and 2.9 hours (Grimler and Roy, 1987).

In each case, the time spent on adult education and participation in political and religious activities was practically non-existent, so that the figures quoted may, roughly speaking, be considered as a fair estimate of the time spent on leisure and rest.

Considerable differences about definitions, about the sample and about the wording of the questions make it impossible to compare the 1966 data with the 1974–75 and 1985 sets of figures, even in terms of general trends. But the two latter sets of data have been adjusted by the INSEE statisticians in order to ensure their comparability with each other. Table 4.3 shows that, in the 1975–85 decade, French adults acquired, on average, 35 minutes more leisure time per day. More detailed tables from INSEE indicate that this trend applied to every category of the urban population, employed or non-employed. The change was mostly due to the fact that, over the same period of time, the length of professional working time decreased by half an hour on average, while as much time as before was devoted to domestic work. The main leisure activities which benefited from the change were, principally, television watching, then sports, shows, entertainment and games such as crossword puzzles, card games, or off-track horse betting (Grimler and Roy, 1987). In 1985–86, the average leisure time for the total sample amounted to 4 hours per day. More than 40% of this time was used for watching television. On average, going out for shows and social life took up more than 1 hour per day. Time devoted to walking and sport totalled 30 minutes, the same as was spent on reading (Grimler and Roy, 1987).

Leisure Participation Surveys (INSEE and Ministry of Culture)

While a systematic comparative analysis of the data in the INSEE and Ministry of Culture surveys is impossible for methodological reasons, it is nevertheless interesting to present results that give a picture of participation in selected leisure activities at various times in the past, as seen by these two institutions, and to observe the trends suggested by these results. The results may be classified under the following headings: (i) leisure activities involving the media; (ii) artistic leisure activities; (iii) social leisure activities; (iv) physical leisure activities; (v) do-it-yourself leisure activities; and (vi) holiday-taking. This classification is an update of the one suggested by Dumazedier and Ripert (1966). Each of these groups is discussed in turn below.

Table 4.3. Changes in time use by adults in urban households, France, 1974 to 1986.

	Hours per day	
Activities*	1974–75	1985–86
Physiological	12.1	11.9
Meals at home	1.7	1.5
Meals away from home	0.5	0.5
Professional & training†	4.0	3.5
Professional	3.4	2.8
Study and training	0.2	0.3
Travel related to professional and training	0.4	4.0
Domestic work	4.4	4.5
Housework (cooking, dishwashing, laundry)	2.7	2.6
Sewing, odd jobs	0.3	0.4
Other travel	0.7	0.8
Caring for people	0.4	0.4
Free time	3.5	4.1
Television	1.4	1.8
Sport	0.1	0.1
Entertainment, going out	0.1	0.1
Games	0.1	0.2

* For each main activity category, only activities showing major change have been selected.
† NB. Figures are averaged over whole week, including Saturday and Sunday; multiply by 7 to obtain estimate of weekly time.
Source: Grimler and Roy, 1987

Activities involving the media
In spite of important discrepancies between the results of the two studies, there are common trends which indicate an increase in daily television viewing, as well as an increase in the reading of magazines, as shown in Table 4.4. There is a striking development of listening to records and cassettes, obviously linked to the growing interest in music which will be reported in the paragraph on artistic leisure activities. A decrease may be observed in the reading of daily newspapers. Contrasting results are presented about radio listening and about going to the movies but, in both cases and for both studies, the range of variation over time is not very significant.

Artistic leisure activities
Looking for trends in arts-based activities, we observe a growing interest in visits to museums (including, probably, the science and technology museums and exhibits, which are becoming more popular, but which cannot be called 'artistic'). Theatre remains restricted to a somewhat limited part of the population, and appears to have remained comparatively stable through the period observed. According to the INSEE survey, the same comment applies to going to the opera or to lyric theatre, while attending the music-hall seems to have lost much of its popularity, possibly because of the numerous television programmes mimicking the music-hall. There is however a growing interest in music, as shown in Table 4.5. The percentage of people actually playing music also increases. Attending classical music concerts or other music (pop, jazz, rock) is gaining popularity, but at a slower pace. Contrasting trends appear in the two studies about visiting artistic exhibitions, as well as visiting castles and historic monuments.

Social leisure activities
Most of the social leisure activities, especially inviting or visiting friends or relatives and belonging to associations, show an increase in participation, as shown in Table 4.6. This is not the case, however, for going to cafés where there is a decrease, or playing music with a group or with friends, and going to fairs, which are static.

Physical leisure activities
Among physical leisure activities, the surveys paid particular attention to formal sport, but provided very little information on more relaxed or informal physical activities. However, it is notable that the INSEE, INSEP, Ministry of Culture and CREDOC surveys have all included a question on walking, with the following highly contrasted results:

INSEE (14+), 1983–1984	11.7%
INSEE (14+), 1988	10.7%
Ministry of Culture, 1973 (in forest or country)	73.0%
Ministry of Culture, 1981	72.1%

Table 4.4. Participation in leisure activities involving the media, France, 1967–1988

	INSEE (14+)			Ministry of Culture (15+)			
		1967 (%)	1988 (%)		1973 (%)	1981 (%)	1988 (%)
Television viewing	Every day or almost every day	51.3	82.6	Every day or almost every day	65.1	69.0	73
Radio listening	Every day or almost every day	67.0	74.2	Every day or almost every day	71.8	72.0	66
Reading a newspaper	Every day	59.7	41.2	Every day	55.1	46.1	43
Reading a magazine or a journal	Regularly	55.7	78.6	Regularly	28.8	–	84
Reading books	At least once per month	32.4	31.4	At least once in last 12 months	69.7	74.0	74
Listening to cassettes or records	Every day or almost every day	8.9	18.7	Every day or almost every day	9.0	9.0	21
Being registered in a library		–	–		13.2	14.3	17
Going to the movies	At least once a month	17.8	18.9	At least once in last 12 months	51.7	49.6	49

Source: see Table 4.1

Table 4.5. Participation in artistic leisure activities, France, 1967–1988.

	INSEE (14+)			Ministry of Culture (15+)			
		1967 (%)	1988 (%)		1973 (%)	1981 (%)	1988 (%)
Going to:							
Theatre	At least once a year	20.9	17.9	At least once in last year	12.1	10	14
Concerts	At least once a year			At least once in last year			
Classical/opera		6.5	7.4		6.9	7	9
Jazz/pop/rock			8.5		6.5	10	13
Museums	At least once in last year	17.8	32.8	At least once in last year	15.4	–	18
Artistic exhibitions	At least once a year	13.8	11.5	At least once in last year	27.4	30	30
Opera	–*		7.4	At least once in last year	18.6	21	23
Lyrics	–			At least once in last year	2.6	2	3
Ballet	–			At least once in last year	4.4	3	3
Music hall	–		15.9	At least once in last year	5.8	5	6
Castles/historic monuments	At least once in last year	70.1	40.3	At least once in last year	11.5	10	10
Playing: Music	Regularly or once in a while	3.7	6.6		31.8	32	28

Source: see Table 4.1

* Classical music or opera

Table 4.6. Participation in social leisure activities, France, 1967–1988.

	INSEE (14+)				Ministry of Culture (15+)		
		1967 (%)	1988 (%)		1973 (%)	1981 (%)	1988 (%)
Going out in the evening	At least once a month	30.2	47.2	At least once a month	15.9	25.8	26
Inviting friends or relatives for a meal	At least once a month	38.6	63.1	At least once a month	–	43.4	50
Visiting friends or relatives for a meal	At least once a month	36.5	60.2	At least once a month	–	Friends 58.2 Relatives 54.3	Once in a while 51 Once in a while 60
Belonging to at least one association	–	11.2	17.2	–	19.5	23.4	27
Dancing	At least 5 or 6 times a year	20.3	29.3	At public dancing hall in last year	25.4	28.0	28
Playing cards or games	Each week or almost each week	13.2	18.2	In the last 12 months often or once in a while	Chess 7.9 Other 41.9	Chess 8.4 Other 45.5 Chance 55.1 Electron. 14.9	–
Going to the café	At least once a week	24.1	17.7	–	–	–	–
Playing music or singing with friends or with a group	–	–	–	In the last 12 months	5.1	5.1	–
Going to a fair	At least once a year	–	29.1	In the last 12 months	47.0	43.0	45

Source: See Table 4.1

Ministry of Culture, 1988 (often or sometimes)   32.0%
INSEP, 1985   24.9%
CREDOC, 1983   24.1%

The main results from leisure participation surveys concerning sports are
to be found in Table 4.7. As was the case for other activities, methodological
obstacles make an accurate diachronic comparison impossible, but trends can
again be observed. In the period between 1967 and 1988, there was an
overall increase of participation in sports: cycling, skiing and gymnastics are
the physical activities for which the rate of participation increased most.
Windsurfing and table tennis also increased, while team sports, except football, became less popular. One should take into account the fact that the
'Clothing survey' in Table 4.7 uses a much wider definition of gymnastics.
Therefore the percentage for gymnastics in the corresponding column should
not be compared to those in the other columns.

Table 4.7. Participation in sports, France, 1967–1988.

	INSEE				INSEP
	1967 (14+) (%)	1974–75 (18+) Time-budget study (urban) (%)	1983–84 (14+) Clothing survey (%)	1988 (14+) (%)	1985 (12–74) (%)
One sport at least	39.0	48.8	43.2	47.7	about 75
Swimming	25.0	29.8	18.1	13.0	22.5
Gymnastics	11.1	12.9	36.0	15.1	26.3
Bowling	13.1	24.9	0.9	–	3.3
Football	5.1	5.5	4.2	3.7	6.8
Basketball	2.7	2.3	1.2	–	1.4
Volleyball	–	–	1.1	–	2.3
Rugby	1.0	1.1	1.1	–	0.6
Ski	4.9	9.7	5.8	16.5	10.0
Athletics	3.5	3.7	–	–	–
Sailing	3.4	5.1	2.5	–	2.3
Tennis	3.0	5.9	8.6	8.3	12.8
Mountain climbing	1.5	–	–	–	1.0
Horseback riding	1.4	2.8	1.0	–	2.8
Martial arts, (judo, karate, etc.)	0.8	1.2*	–	1.3†	0.7
Cycling	–	–	9.2	13.8	15.4
Running	–	–	3.6	4.8	12.7
Windsurfing	–	–	1.1	1.8	4.0
Table tennis	–	–	1.0	3.1	3.9
Body building/Weight lifting	–	–	–	2.6	0.8

Source: see Table 4.1
* Including karate.
† Martial arts.

'Practical' leisure activities
The leisure participation surveys give limited information on 'practical' leisure activities, such as do-it-yourself, gardening, hunting, and fishing, probably because of the reluctance of many researchers to include them within their definition of leisure. We know, however, from INSEE that in 1987–88, 23.5% of male respondents stated that they repaired their car for pleasure once in a while, while 67.2% of female respondents said that they enjoyed sewing once in a while. This seems to show that, at least for some people, 'do-it-yourself' or 'practical' activities are indeed a form of leisure.

From the trends indicated in Table 4.8, we see that do-it-yourself activities are increasing and that gardening is decreasing. Hunting and fishing are both reported as decreasing but it must be noted that the INSEE figures for hunting and fishing given for 1967 and 1987–88 are not comparable because the seasons of reference are not the same.

Holiday taking
Since 1965, as mentioned before, the yearly results concerning holiday taking have been collected on a strictly comparable basis. Also, data from the previous INSEE surveys have been adjusted by statisticians, so that they could be compared with the series starting in 1965. The surveys therefore provide a continuous record of the proportion of the population taking a summer holiday away from home (for at least four consecutive days) and, as Table 4.9 shows, there has been a dramatic increase, from 30% in 1951 to 55% in 1992.

Table 4.8. Participation in 'practical' leisure activities, France, 1967–1988.

		INSEE (14+)		Ministry of Culture (15+)				INSEP (12–74)
		1967 (%)	1987–88 (%)		1973 (%)	1981 (%)	1988 (%)	1985 (%)
Do-it-yourself	Once or twice a week	16.0 6.7*	20.5	In the last 12 months	37.5	58.5	–	–
Hunting	Regularly or not during the hunting season	6.9	4.5	In the last 12 months	9.2	6.5	5.0	2.2
Fishing	Regularly or not during the fishing season	17.8	11.0	In the last 12 months	20.0	17.0	16.0	2.2
Gardening	Every day or almost every day in 'good' season	20.2	18.2	In the last 12 months	54.1	–	38.0	–

* For pleasure

Conclusion

In summary, important leisure trends in the period 1967–1987 are the growth of activities involving the media (especially television) and the popularity of music; trends which can be described as an individualization of the

Table 4.9. Holiday taking, France, 1951–1992.*

	Holidays over the whole year (%)	Summer holidays (%)	Winter holidays (%)	Winter sports (%)
1951	–	30.0†	–	–
1957	–	39.0†	–	–
1961	–	39.0†	–	–
1964	43.6	41.7†	–	–
1965	45.0	41.0	–	–
1966	52.5	41.7	–	–
1967	54.0	42.6	–	–
1968	55.3	41.6	–	–
1969	54.3	42.7	–	–
1970	56.0	44.6	–	–
1971	56.2	46.0	–	–
1972	57.2	–	–	–
1973	57.8	–	–	–
1974	57.3	–	–	–
1975	57.5	50.2	17.1	–
1976	54.0	51.6	18.1	–
1977	55.3	50.7	17.9	–
1978	54.3	51.7	20.6	–
1979	56.0	53.3	22.1	–
1980	56.2	53.8	22.7	–
1981	57.2	54.3	23.8	–
1982	–	54.5	24.6	–
1983	–	–	–	–
1984	57.3	53.1	26.2	10.0
1985	57.5	53.8	24.9	8.8
1986	58.2	54.1	27.1	9.6
1987	58.5	54.2	28.2	8.8
1988	–	55.5	28.2	8.8
1989	60.7	56.5	27.3	7.9
1990	59.1	55.1	26.7	7.1
1991	59.8	56.6	26.3	8.4
1992	60.0	55.3	28.9	8.8

– Not available; * for at least 4 consecutive days; † estimates to ensure data comparability.
Sources: INSEE: *Donnees Sociales*, 1973, 1987; Bertrand, 1984; Christine, 1987; Memento du Tourisme, 1993.

relationship between people and culture (or 'individuation' as Rojek (1985, p. 19) calls it). Television competes with other leisure activities like movies or reading newspapers and books, with a decline in reading in the population except for the highly educated. More than half of the population ignores leisure activities which are often seen as characteristic of the French 'traditional culture', such as going to the theatre, to a concert of classical music, to a dance performance, to an art exhibition or to museums and historic monuments (in 1990 more French people visited an industrial site or a technological museum (67%) than art museums (57%) – Ficquelmont, 1991; Samuel, 1993).

The overall increase in social leisure activities is confirmed by other studies, which link this intensification to a need for 'belonging' in a society which is often felt as alienating, and to a preference for achieving this aim through activities oriented towards self-realization (leisure) rather than through political or religious participation, which is another option for the use of free time (Hatchuel and Valadier, 1992).

Concerning 'practical' leisure, the increase of do-it-yourself activities may be a response to the growing cost of repair and construction work done by specialists. The decrease in gardening which appears from surveys seems in contradiction to the development of 'garden centres' and with the popularity of books, magazines, radio and television programmes dealing with the subject.[2] The decrease in hunting and fishing may be a consequence of the action of environmentalists and of the subsequent strengthening of hunting and fishing laws; the growing pollution of the waters and the disappearance of valuable species may also play a part in the case of fishing.

There has been an increase in participation in sport in France, and the range of sporting activities has widened. The favourite sports are activities which can be performed all year round, at any age, either alone or with the family (Garrigues, 1989). It may be said that the 1980s were characterized by a rediscovery of the body and by a growing awareness of its importance as a health factor.

Finally, holiday-taking has been a growing phenomenon since the 1950s (with the increase of paid vacation time), but has remained static since the end of the 1980s, probably as a result of the economic crises and of the rise of unemployment.

LEISURE PARTICIPATION AND SOCIOECONOMIC VARIABLES

As mentioned above, the results of both the INSEE and the Ministry of Culture surveys are tabulated according to a number of sociocultural variables. Results of these tabulations are not comparable because of differences in the choice of categories for the analysis of responses. But, as in overall participation rates, trends may be observed. For the sake of brevity and clarity, comments will only be made about the leisure activities showing significant differences in participation between the categories of population under observation. The relationships between leisure and gender, age, occupation, size of community and level of education are discussed in turn.

Gender

As indicated in the discussion of time-budget studies above, working women have less leisure than working men, respectively 2.9 hours and 3.7 hours a day in 1985–86 (Grimler and Roy, 1987). But women not in paid employment have the highest average amount of daily free time: 4.1 hours in 1985–86. This difference is especially apparent when it comes to television viewing (Grimler and Roy, 1987). As shown in Table 4.10, women are somewhat less inclined to go out in the evening, to go to a café and to visit commercial shows. Their membership of associations is lower. Among practical leisure activities, they do very little repair work but, as indicated above, the traditional activity of sewing gives pleasure to many. Finally, women continue to be less active in sports than men, although their participation has increased in the last 20 years (Garrigues, 1989).

Age

The 1985 Time Budget Study information on daily average free time according to age is presented in Table 4.11. As may be expected, at both ends of the age scale, the non-employed people, either below 18 or above 65, benefit from a longer period of free time (Grimler and Roy, 1987).

It is not surprising either to observe that young people aged 14–24 are characterized by the highest rate of participation in many leisure activities (see Table 4.12), including: radio listening (81.4%), playing music (14.9%), going to the theatre (20.1%), going to the movies (44.8%), going to the café (30.6%), dancing (58.3%), going out at night (67.1%), playing card games (24.3%), visiting museums (43.7%), and collecting (38.6%). They also have the highest rate for doing at least one sport (27.9%), for doing athletics (21.2% of the 14–18 year olds), swimming (54.5%) and team sports like football, soccer (25.9%), basketball (17.1%) or rugby (5.2%) (Garrigues, 1988, pp. 43, 55).

The leisure behaviour of elderly people (60 and over) is characterized by the highest rate of television viewing (91.6%), of daily newspaper reading

(56.5%), of gardening (33.3%) and by a low rate of participation in sports and in activities carried out away from home. A most striking discovery!

Occupation

Leisure participation related to occupation is shown in Table 4.13. As can be expected, managers and professionals are characterized by a high rate of participation in many leisure activities, such as magazine reading (92.6%), book reading (57.1%), going to the movies (37.8%), going to the theatre (46%), going to museums (63.9%), visiting castles and historic monuments

Table 4.10. Leisure participation by gender, France, 1987–88.

	Males (%)	Females (%)	Total (%)
TV every day or almost every day	82.7	82.4	82.6
Newspaper reading every day or almost every day	44.7	38.0	41.2
Reading a journal or magazine regularly	75.1	81.8	78.6
Reading at least one book per month	26.8	35.5	31.4
Going to the movies at least once per month	19.5	18.3	18.9
Going to the theatre at least once a year	16.0	19.7	17.9
Going to a sport show at least 5 times a year	14.6	3.2	8.6
Going to a fair or commercial show in the last year	59.8	50.1	54.7
Going to a museum in the last year	33.3	31.9	32.6
Going to a castle or historical monument in last year	41.2	39.5	40.3
Going out in the evening at least once a month	53.2	42.5	47.6
Going to the café at least once a week	26.2	9.8	17.7
Going to a restaurant at least once a month	27.0	22.9	24.6
Inviting friends or relatives at least once a month for a meal	64.0	62.2	63.1
Visiting friends or relatives at least once a month for a meal	61.5	59.0	60.2
Being a member of one association at least	20.8	13.9	17.2
Repairing a car for pleasure once in a while	23.5	1.3	11.9
Sewing for pleasure once in a while	5.2	67.2	37.6
Gardening	20.0	17.7	18.8
Radio listening every day or almost every day	76.5	72.1	74.2
Dancing at least 5 or 6 times a year	29.6	29.1	29.3
Playing card games or others every week or almost	20.2	16.3	18.2
Playing music regularly or once in a while	7.3	5.9	6.6
Making a collection	23.5	21.5	22.5
Doing one sport (boules excluded)	53.4	42.5	47.7

Source: Dumontier and Valdelievre, 1989.

Table 4.11. Daily free time by age, gender and employment status, France, 1985–1986. |

| | Free time, hours per day | | | |
| | Males | | Females | |
Age	Employed	Non-employed	Employed	Non-employed
15–17 years	–	4.95	–	4.38
18–24 years	4.03	5.10	3.35	4.18
25–54 years	3.50	5.62	2.65	3.70
55–64 years	3.48	5.73	2.72	4.32
65–74 years	–	5.98	–	4.88
75 years and over	–	5.88	–	4.73
Overall average		4.00		

Source: Grimler and Roy, 1987.

(71.7%), going out in the evening (71.4%), going to restaurants (48.9%) inviting (82.3%) or visiting (81%) friends or relatives for a meal, belonging to associations (27.3%), playing music (14.3%) and collecting (35.7%). Concerning sports, they have the highest rate of participation in windsurfing only (37%),[3] although they are active in many other sporting activities, such as gym (24.8%), horseback riding (25.7%), walking and hiking (22%), swimming (23.9%), skiing (31%), tennis (30.2%) and volleyball (26.9%) (Garrigues, 1988, p. 64; see also Table 4.10).

Manual workers are especially active in sports: they have the highest rate of participation for doing at least one sport (30.3%), and specifically for basketball (38.2%), cycling (30.2%), football (47.9%), walking and hiking (26.5%), swimming (25.1%), bowling (45.8%), and volleyball (20.5%) (Garrigues, 1988, p. 64).

Skilled personnel are characterized by high rates of participation for visits to commercial shows (72.6%), radio listening (82.9%), and mostly in sports, such as running/jogging (31.6%), horseback riding (28.7%), gym (27.6%), swimming (26.3%), table tennis (29.1%), skiing (33.6%) and tennis (30.8%) (Garrigues, 1988, p. 64).

Size of Community

From the 1985 Time Budget Study, we know that rural dwellers spend less time than urban dwellers in television viewing, on social events and on attending shows. Young rural men, employed or not, spend more time hunting and fishing. More generally, outdoor recreation takes up as much time in rural areas as in urban areas. Altogether, however, the average daily free time available is 30 minutes less for rural than for urban dwellers (Grimler and Roy, 1987).

People living in rural areas are characterized by a higher rate of participation in associations and in gardening (Table 4.14), while their participation in sports is low: only 6.8% play at least one sport. Their rate of participation is the lowest for the following sports: basketball (7.3%), running/jogging (4.9%), cycling (8.2%), horseback riding (2.7%), gym (5.2%), walking and hiking (4.4%), swimming (3.7%), table tennis (4.2%), skiing (3.5%), tennis (5.7%), volleyball (3.3%) and windsurfing (3.1%) (Garrigues, 1988, p. 76).

From Table 4.14 and Table 4.10 it appears that people living in cities of 100,000 inhabitants and over (Paris not included) have the highest rate of participation for going to commercial shows, for playing card games and for some sports, namely: basketball (32.3%), jogging/running (30.5%), cycling (26.4%), horseback riding (30.5%), football (soccer) (27.8%), walking or

Table 4.12. Leisure participation by age, France, 1987–88.

	14–24 years (%)	29–35 years (%)	40–59 years (%)	Over 60 years (%)	Total (%)
TV every day or almost every day	78.0	77.4	83.6	91.6	82.6
Newspaper reading every day or almost every day	24.8	35.3	46.5	56.5	41.2
Reading a journal or magazine regularly	83.2	83.8	76.2	71.2	78.6
Reading at least one book a month	41.6	34.5	25.6	25.5	31.4
Going to the movies at least once a month	44.8	20.4	10.5	4.4	18.9
Going to the theatre at least once a year	20.1	18.3	18.9	14.4	17.9
Going to a sport show at least 5 times a year	11.8	10.5	8.4	4.0	8.6
Going to a museum in the last year	43.7	35.8	30.8	21.1	32.6
Going to a castle or historical monument in the last year	43.3	45.1	43.0	28.6	40.3
Going out in the evening at least once a month	67.1	63.7	43.4	16.2	47.6
Going to the cafe at least once a week	30.6	20.0	14.6	7.2	17.7
Going to the restaurant at least once a month	27.3	32.6	23.2	14.2	24.6
Inviting friends or relatives at least once a month for a meal	64.8	71.6	63.9	50.2	63.1
Visiting friends or relatives at least once a month for a meal	62.0	74.8	56.3	45.4	60.2
Being a member of one association at least	11.8	17.5	20.1	18.1	17.2
Repairing a car for pleasure once in a while	10.6	16.7	13.3	5.5	11.9
Sewing for pleasure once in a while	29.0	36.9	40.1	43.2	37.6
Gardening	3.7	13.3	25.2	33.3	18.8
Radio listening every day or almost	81.4	78.2	74.1	63.2	74.2
Dancing at least 5 or 6 times a year	58.3	34.5	21.5	7.1	29.3
Playing card games or others every week or almost	24.3	15.8	13.3	21.5	18.2
Playing music regularly or once in a while	14.9	6.9	3.8	2.1	6.6
Making a collection	38.6	23.0	19.7	11.1	22.5

Source: Dumontier and Valdelievre, 1989.

Table 4.13. Leisure participation by occupation, France, 1987–88.

	Farmers		Independent workers (%)	Managers professionals (%)	Skilled personnel (%)	Employees (%)	Workers (manual) (%)	Services (%)	Others (%)	Non-working (%)	Total (%)
	Owners (%)	Wage earners (%)									
TV every day or almost every day	82.5	88.5	78.7	71.3	73.4	80.4	83.9	80.1	79.1	89.4	82.6
Newspaper reading every day or almost every day	48.0	31.6	45.3	39.5	33.4	35.9	33.3	32.0	31.5	52.5	41.2
Reading a journal or magazine regularly	70.3	62.2	84.5	92.6	90.7	86.8	74.7	75.7	92.9	71.6	78.6
Reading at least one book a month	10.9	21.4	29.6	57.1	45.9	34.9	24.2	29.1	44.8	26.7	31.4
Going to the movies at least once a month	10.6	17.4	21.8	37.8	28.7	27.8	15.5	31.1	25.1	10.0	18.9
Going to the theatre at least once a month	8.3	5.3	19.6	46.0	31.2	19.0	8.6	16.1	21.1	14.0	17.9
Going to a sport show at least 5 times a year	9.8	13.7	10.9	7.2	9.6	11.6	11.5	6.6	8.8	5.1	8.6
Going to a museum in the last year	17.9	16.5	32.1	63.9	52.2	32.6	25.9	34.2	47.3	23.7	32.6
Visiting a castle or historic monument in the last year	24.6	26.4	42.9	71.7	57.7	45.9	33.9	37.8	57.4	30.2	40.3
Going out in the evening at least once a month	39.9	47.1	55.6	71.4	68.8	61.5	52.1	56.6	57.3	24.5	47.6
Going to a café at least once a week	10.0	17.1	20.4	21.8	22.1	25.1	20.7	22.8	17.2	10.8	17.7
Going to a restaurant at least once a month	5.9	10.8	34.0	48.9	40.4	31.3	18.4	27.8	30.8	16.3	24.6
Inviting friends/rels for meal at least once a month	58.0	60.3	68.9	82.3	76.1	67.3	61.5	59.7	69.7	52.8	63.1

Visiting friends/rels for meal at least once a month	44.4	55.0	61.8	81.0	77.4	71.8	60.6	63.8	69.0	46.2	60.2
Being a member of at least 1 association	23.6	8.5	16.7	27.3	23.3	15.5	13.0	12.7	13.8	16.3	17.2
Repairing car for pleasure once in a while	12.0	15.9	10.0	8.0	12.4	11.0	18.3	11.4	16.2	7.5	11.9
Sewing for pleasure once in a while	33.8	38.1	32.3	30.5	32.5	38.4	37.1	42.8	36.8	43.0	37.6
Gardening	23.8	27.3	11.4	8.8	10.2	11.3	17.2	11.6	11.7	29.3	18.8
Listening to radio (almost) every day	69.1	77.7	75.9	81.0	82.9	80.2	75.8	80.6	77.2	66.1	74.2
Dancing at least 5/6 times a year	32.7	32.4	39.7	32.0	39.5	39.8	34.7	37.7	36.4	14.3	29.3
Playing card/other games (almost) every week	12.2	11.9	16.3	14.6	14.5	15.6	19.5	15.8	18.9	21.5	18.2
Playing music regularly or once in a while	4.5	4.6	8.0	14.3	11.7	5.9	5.4	8.8	8.6	3.3	6.6
Collecting	11.8	9.7	26.4	35.7	32.2	25.2	23.0	32.7	33.2	13.6	22.5

Source: Dumontier and Valdelievre, 1989.

Table 4.14. Leisure participation by size of community of residence, France, 1987–88.

	Population					
	Rural (%)	< 20,000 (%)	20,000 to 100,000 (%)	> 100,000 except Paris (%)	Paris area (%)	Total (%)
TV viewing every day or almost every day	83.5	84.3	81.7	81.4	82.2	82.6
Newspaper reading every day or almost	45.1	44.3	43.0	43.3	26.7	41.2
Reading a journal or magazine regularly	72.6	77.8	81.7	81.1	83.0	78.6
Reading at least one book a month	23.4	27.0	32.8	35.5	41.2	31.4
Going to the movies at least once a month	11.2	15.2	19.8	20.4	32.1	18.9
Going to the theatre at least once a year	9.5	13.5	18.0	18.2	35.9	17.9
Going to a sport show at least 5 or 6 times a year	9.4	9.9	10.5	9.1	3.8	8.6
Going to a museum in the last year	23.4	28.9	31.1	35.7	47.6	32.6
To a castle or historical monument in the last year	33.0	37.8	40.3	40.0	55.2	40.3
Going out in the evening at least once a month	39.9	46.1	46.5	51.0	57.2	47.6
Going to the café at least once a week	13.2	16.3	17.4	16.6	28.3	17.7
Going to a restaurant at least once a month	13.8	19.9	21.9	28.6	42.7	24.6
Entertaining friends/relatives once a month for a meal	60.2	58.7	59.9	65.8	70.7	63.1
Visiting friends/relatives once a month for a meal	51.4	58.1	59.3	64.5	70.3	60.2
Being a member of one association at least	20.3	18.3	16.4	15.9	13.8	17.2
Repairing a car for pleasure once in a while	13.3	12.4	11.8	11.6	9.5	11.9
Sewing for pleasure once in a while	37.7	39.0	40.2	37.3	34.6	37.6
Gardening	31.9	22.3	16.1	11.6	7.6	18.8
Radio listening every day or almost every day	72.1	72.2	75.7	74.0	78.9	74.2
Dancing at least 5 or 6 times a year	29.1	29.5	29.7	28.6	30.4	29.3
Playing cards/other games every week or almost	17.2	18.6	18.7	19.6	16.6	18.2
Playing music regularly or once in a while	5.6	5.9	5.7	7.7	7.5	6.6
Making a collection	16.0	20.5	26.2	24.1	29.5	22.5

Source: Dumontier and Valdelievre, 1989.

hiking (32.9%), swimming (32.6%), bowling (27.4%), table tennis (24.5%), skiing (33.7%), tennis (34.5%), volleyball (39.5%) and windsurfing (37.1%). Altogether, 30.6% do at least one sport (Garrigues, 1988, p. 76).

People living in Paris or in the Paris area are characterized by a high rate of participation in the following activities: reading magazines (83.0%), reading books (41.2%), going to the movies (32.1%), going to the theatre (35.9%), going to museums (47.6%), visiting castles or historic monuments (55.2%), going out in the evening (57.2%), going to cafés (28.3%) and restaurants (42.7%), entertaining (70.7%) and visiting (70.3%) relatives or friends for a meal, radio listening (78.9%), dancing (30.4%) and collecting (29.5%). Among sports, their rates of participation in swimming (27.4%) and in table tennis (25.9%), are high (Garrigues, 1988, p. 76).

Level of Education

Participation surveys confirm common observations: participation in many leisure activities is highly dependent upon educational variables (see Table 4.15). For example, university graduates have the highest rate of participation in: reading magazines (92.3%), reading books (58.8%), going to the movies (36.7%), going to the theatre (41.9%), visiting commercial shows (71%), museums (60.9%) and castles and historic monuments (66%), going out in the evening (72.5%), visiting restaurants (49.3%), entertaining (77.8%) or visiting (79.5%) friends and relatives for a meal, belonging to associations (26.4%), playing music (11.8%) and collecting (30.6%).

The situation is, however, strikingly different in the field of sports: only 11.8% of people with a university degree play at least one sport. Even though 22.7% of this group go skiing, 20.6% play tennis and 28.8% go windsurfing, other educational categories have higher rates of participation in these and other sports.

People having no degree at all have a high average rate of participation in sports, especially team sports: 21.3% do at least one sport, 42.7% play basketball, 34.3% play football, 28.3% play volleyball, 36.7% go bowling and 22% play table tennis (Garrigues, 1988, p. 71).

In brief, leisure participation surveys show that, on the whole, being a young man, living in the Paris area, being a university graduate and having a managerial or professional occupation are determining factors in an active participation in most leisure activities.

CONCLUSION

As may be gathered from the preceding sections, the problems usually listed as plaguing survey methodology in general (Cushman and Veal, 1993, p. 212)

Table 4.15. Leisure participation by level of education, France, 1987–88.

	No degree (%)	Primary school (%)	Technical certificate (%)	High school & university (%)	Total (%)
TV every day or almost every day	86.7	89.1	81.0	70.7	82.6
Newspaper reading every day or almost	40.0	45.8	37.9	42.0	41.2
Reading a journal or magazine regularly	61.4	78.2	85.7	92.3	78.6
Reading at least one book a month	15.2	26.0	32.8	58.8	31.4
Going to the movies at least once a month	9.3	10.2	23.4	36.7	18.9
Going to the theatre at least once a year	6.0	12.3	17.9	41.9	17.9
Going to a sport show at least 5 times a year	5.9	8.1	11.8	8.1	8.6
Going to a museum in the last year	15.3	25.5	35.9	60.9	32.6
Going to a castle or historical monument in the last year	21.8	35.5	44.5	66.0	40.3
Going out in the evening at least once a month	29.7	32.8	60.4	72.5	47.6
Going to the café at least once a week	14.4	11.9	22.7	21.9	17.7
Going to a restaurant at least once a month	12.2	15.4	27.7	49.3	24.6
Inviting friends/relatives at least once a month for a meal	51.3	56.7	69.6	77.8	63.1
Visiting friends/relatives at least once a month for a meal	48.1	50.1	67.1	79.5	60.2
Being a member of one association at least	10.0	16.8	18.1	26.4	17.2
Repairing a car for pleasure once in a while	9.7	9.8	16.7	10.2	11.9
Sewing for pleasure once in a while	41.1	41.9	33.1	34.2	37.6
Gardening	23.2	26.2	13.3	11.3	18.8
Radio listening every day or almost every day	65.5	71.7	80.4	80.1	74.2
Dancing at least 5 or 6 times a year	18.8	22.9	42.0	32.7	29.3
Playing cards or other games every week or almost	18.4	19.4	18.3	16.0	18.2
Playing music regularly or once in a while	3.0	5.4	7.4	11.8	6.6
Making a collection	12.2	19.6	29.0	30.6	22.5

Source: Dumontier and Valdelievre, 1989.

are indeed present in the French leisure participation surveys. The most important among these problems relate to the scope of leisure, to sampling, to the choice of the period of reference and to the choice of categories for the analysis of responses.

As to the scope of leisure, this problem goes back to the old quarrel about the definition of leisure which has been going on for decades (Dumazedier, 1974, pp. 88–134) and which, of course, is strongly associated with the listing of leisure activities. The French surveys tend to define leisure in a broad sense, including a wide range of home-based, social and entertainment activities as well as outdoor activities. But each survey is focused according to the central interests of its sponsoring institution (for example the Ministry of Culture surveys emphasize artistic leisure activities). Some leisure activities are, however, in practice excluded from French surveys: for example cigarette or pipe smoking was only investigated in the first INSEE, 1967, study (Debreu, 1973, p. 6); going to cafés is included but no mention of drinking is to be found; there is no reference whatsoever to sexual activity, which obviously plays a role in leisure activities, *a fortiori* for drugs.

Sampling, a matter of importance, is done in different ways in practically every survey, with a particularly striking lack of uniformity in the age ranges of the respondents.

The choice of the period of reference varies from one survey to another and even, sometimes, within the same survey for different questions. The most usual procedure is to ask the respondents to indicate their participation over the course of the previous year. But, in some cases, a shorter reference period, namely the month before the interview, has been used, which makes it difficult to synthesize the leisure behaviour of the respondents and which may introduce a bias in the case of seasonal leisure activities.

Finally, there is a lack of uniformity in the choice of the sociocultural categories used in the analysis of responses: age groups vary from one survey to another; the same goes for occupational groups and educational groups. The only sociocultural variable which is uniformly treated is gender!

It may be hoped that these shortcomings will be remedied in the future,[4] so that leisure participation surveys will increase their potential as a resource for leisure research. Of course, it may be argued that we do not need large participation surveys to know that leisure participation changes over time and varies according to gender, age, occupation, level of education and place of residence, but the basic principle of the scientific method reminds us that any statement has to be tested against the evidence before we can be sure of its validity and surveys are one means of providing such evidence.

NOTES

1. Making comparison painful, if not impossible. But this is called progress!
2. Maybe buying, reading, listening and viewing take up so much time that very little is left for gardening proper!
3. I have recently described the rising vogue of 'gliding sports' (Samuel, 1993). As in many other cases, the élite seem to set the example. Their 37% rate of participation in these sports is closely followed by the 27.1% of the skilled employees, the figures for the other occupational groups being much lower.
4. But the comparison with previous surveys might then become even more difficult than it is today.

REFERENCES

Bertrand, M. (1984) Les vacances. *Données Sociales*, pp. 253–257.

Chombart de Lauwe, P.H. (1958) La vie quotidienne des familles ouvrieres. *Recherches sur les Comportements Sociaux de Consommation* CNRS, Paris.

Christine, M. (1987) Les vacances. *Données Sociales*, p. 382.

Centre de Recherche et de Documentation sur la Consommation (CREDOC) (1983) *Le système d'Enquêtes sur les Conditions de Vie et Aspirations des Français: 1978–1981, Phase 4, Vol. 7: Le Temps Libre, les Vacances.* CREDOC, Paris.

Cushman, G. and Veal, A.J. (1993) The new generation of leisure surveys. Implication for research on everyday life. *Loisir et Société* 16 (1), pp. 211–220.

Debreu, P. (1973) *Les Comportements de Loisir des Français*, Coll. INSEE, No.102, Serie M, No.25, August (Also: *Economie et Statistique*, No.51, December).

Documentation Française (1988) *L'enquête Emploi du Temps.* Documentation Française, Paris.

Donnat, O. and Cogneau, D. (1990) *Les Pratiques Culturelles des Français.* Documentation Française, Paris.

Dumazedier, J. (1962) *Vers une Civilisation du Loisir?* Le Seuil, Paris.

Dumazedier, J. (1974) *Sociologie Empirique du Loisir.* Le Seuil, Paris.

Dumazedier, J. and Ripert, A. (1966) *Loisir et Culture.* Le Seuil, Paris, pp. 319–325.

Dumontier, F. and Valdelievre, H. (1989) *Les Pratiques de Loisir: Vingt Ans Après: 1967/1987–88.* INSEE Resultats No. 13. Institut National de Statistique et d'Études Economiques, Paris.

Erlinger, P., la Louveau, C. and Metoudi, M. (1985) *Pratiques Sportives des Français.* Ministère de Jeunesse et Sports, Paris.

Ficquelmont, G.M. de (1991) La visite d'entreprise, un loisir culturel. *Espaces* 107, Dec–Jan., 31–34.

Fourastie, J. (1951) *Machinisme et Bien-etre.* Editions de Minuit, Paris.

Garrigues, P. (1988) *Evolution de la Pratique Sportive des Français de 1957 à 1984.* Coll. de l'INSEE, 134 M., October. Institut National de Statistique et d'Études Economiques, Paris.

Garrigues, P. (1989) Une France un peu plus sportive qu'il y a 20 ans . . . grace aux femmes. *Economie et Statistique*, No. 224, September.

Girard, A. (1958) Le budget-temps de la femme mariée dans les agglomerations urbaines. *Populations* No. 4, October–December.

Girard, A. and Bastide, H. (1959) Le budget-temps de la femme mariée à la campagne. *Populations* Avril–Juin.

Goguel, C. (1965) Les vacances de Français en 1964. *Economie et Statistiques*, No. 5, Juin.

Goguel, C. (1966) Recherche comparative internationale sur les budgets-temps. *Études et Conjoncture* No.9, September.

Gounot, Ph. (1958) Les vacances des Français en 1957. *Études et Conjoncture* No.7, Juillet.

Gounot, Ph. (1961) Les vacances des Français en 1961. *Études et Conjoncture*, 17eme année, Mai, 413–434.

Grimler, G. and Roy, C. (1987) *Time Use in 1985–86, Paris, Premiers Resultats*, INSEE Resultats No.100, October. Institut National de Statistique et d'Études Economiques, Paris.

Hatchuel, G. and Valadier, J.L. (1992) *Les Grands Courants d'Opinion et de Perception en France de la Fin des Années 70 au Début des Années 90*. Coll. des Rapports, No.116, Mars. CREDOC, Paris.

Huet, M.T., Lemel, Y. and Roy, C. (1978) *Les Emplois du Temps des Citadins*. Resultats Provisoires de l'enquête emplois du temps 1974–1975. INSEE, Paris.

Institut National de Statistique et d'Études Economiques (INSEE) (1973) Les vacances 1951–1970. *Données Sociales*. INSEE, Paris, 87–89

Institut National de Statistique et d'Études Economiques (INSEE) (1974) Les vacances. *Coll. de l'INSEE*, M. 39. INSEE, Paris, pp. 177–179.

Institut National de Statistique et d'Études Economiques (INSEE) (1987) *Données Sociales*. INSEE, Paris.

Lemel, Y. (1972) Elements sur les budgets-temps des citadins. *Economie et Statistiques* 33, Avril.

Lemel, Y. (1974) Les budget-temps des citadins. *Coll. de l'INSEE*, M. 33, Octobre INSEE, Paris.

Le Play, F. (1877) Les ouvriers Européens. Paris, *Mame*, 6 Vol. (2eme edn.).

Leroux, P. (1970a) Le comportement de loisirs des Français. Paris, *Coll.INSEE*, 24, M, 2, Juillet.

Leroux, P. (1970b) Les loisirs des Français. *Economie et Statistiques*, Mai.

Mémento du Tourisme (1993) *Transports et Tourisme*. (Paris, Min. Equipment), Direction du Tourisme, September.

Ministère de la Culture (1974) Pratiques culturelles des Français. *Service des Études et Recherches*, 2 Vol. La Documentation Française, Paris.

Ministère de la Culture (1982) *Pratiques Culturelles des Français, Description Socio-démographique: Evolution 1973–1981*. Dalloz, Paris.

Ministère de la Culture (1990) *Nouvelle Enquète sur les Pratiques Culturelles des Français en 1989*. La Documentation Française, Paris.

Raymond, H. (1968) Loisir, travail, culture 1955–1968. In: *La Sociologie du Loisir* No. Spécial de *Current Sociology* 16 (1) 5.

Rojek, C. (1985) *Capitalism and Leisure Theory*. Tavistock, London.

Samuel, N. (1993) Leisure research for a world in turmoil. Paper to the World leisure and Recreation Association Congress *Leisure, Tourism and Environment*, University of Rajahstan, Jaipur, India, December 5–10.

Samuel, N. and Romer, M. (1984) *Le Temps Libre: Un Temps Social*. Klincksieck – Les Méridiens, Paris.

Société Française d'Études et de Sondages (SOFRES) (Annual since 1990) *Le Suivi de la Demande Touristique des Français*. SOFRES, Paris.

Sondages (1947) *Distraction et Culture en France*, November. IFOP, Paris.

Sondages (1949) *Les Loisirs*, January. IFOP, Paris.

Sondages (1962) *L'étalement des Congés*. IFOP, Paris.

Sondages (1964) *Les vacances en 1963*, No. 2. IFOP, Paris.

Stoetzel, J. (1948) Une étude du budget-temps de la jeunesse dans les agglomérations urbaines. *Populations*, No.1, Jan–Mars.

Szalai, A. *et al.* (1972) *The Use of Time: Daily Activities of Urban and Suburban Populations in 12 Countries*. Mouton, The Hague.

5 Germany

WALTER TOKARSKI AND HARALD MICHELS

BACKGROUND

The only broad study of leisure participation in Germany on a national base is the so-called EMNID Leisure Survey (ELS). The EMNID Institute in Bielefeld is a commercial organization and has conducted this survey every year for more than 25 years. The institute was and is responsible for a large number of surveys in the field of leisure in the broadest sense, including studies on sport images, sport participation, tourism, consumer behaviour and media analysis. The EMNID Institute works primarily for industrial and political clients. The ELS is a nationwide leisure survey conducted completely independently, the results being available to anyone willing to pay some DM5000 for a copy of the annual report or who is interested in having questions included in the questionnaire, for a fee.

No other research institute in Germany produces similar, regular, nation-wide leisure participation surveys although a number of research institutes sometimes conduct one-off leisure studies, including the Allensbach Institute at the Bodensee, the Infratest Institute and the Institute for Leisure Economy in Munich, and the BAT Leisure Research Institute (part of the British-American Tobacco Industries company) in Hamburg. These institutes work on a smaller scale and conduct occasional *ad hoc*, cross-section studies on leisure.

© 1996 CAB INTERNATIONAL. *World Leisure Participation* (eds G. Cushman, A.J. Veal and J. Zuzanek)

THE EMNID LEISURE SURVEY (ELS)

Method

This chapter is based on the results of the ELS 1991, which is the latest available report that can be used for research purposes outside the EMNID institute. The aim of the survey was to present an analysis of the leisure situation in general and of leisure participation rates for certain leisure activities. This was achieved for the first time in West *and* East Germany, one year after the re-unification of the two German states. The interviews took place between 1 August and 13 September 1991, in 572 selected areas, with dwellings/respondents being selected by use of the 'random route' method. The sample included 3069 individuals aged 14 years and over in total, with 2069 from West Germany and 1000 from East Germany (the former German Democratic Republic).

Results

The 1991 ELS report includes a West–East perspective, some general information about the importance of leisure in Germany and specific information about individual leisure activities and participation rates. The report does not include details of all of the 81 activities listed in the ELS questionnaire. Furthermore the report does not include participation tables relating to age, gender, occupation/class, region, income, household data and other data collected in the survey but gives commentary on this if necessary.

The importance of leisure in Germany

Among eight 'life spheres' leisure and holidays are ranked as the third and fourth most important, respectively, after health and family, but ahead of work and politics, as shown in Figure 5.1. Sport is considered separately from leisure and is ranked seventh overall, ahead of religion. The differences between West and East Germany are not significant. In general it can be said that leisure is somewhat more important for men than for women. The younger people are the more important leisure is for them, with a marked decrease in the importance of leisure for people over 60, as shown in Figure 5.2.

The survey asked respondents about the amount of time available for leisure during the week and on weekends. On average people reported having 5.1 hours a day for leisure during the week and between 7.3 hours (Saturdays) and 8.4 hours (Sundays) at weekends, as shown in Figure 5.3. Some differences were revealed between men and women and between different socioeconomic groups. During the week men and women, on average devote

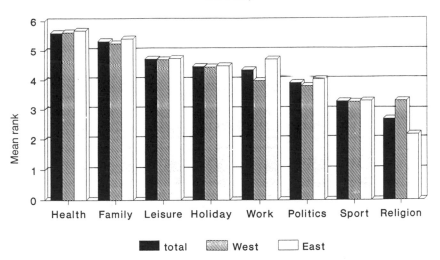

Fig. 5.1. The importance of life spheres in West and East Germany, 1991 (mean ranks).

a similar amount of time to leisure activities, but at weekends men have on average more time. People with lower income seem to have more leisure time than people with higher income over all the days of the week.

In the ELS people were asked if their time budget had changed during the last 5 years. One-quarter said that there had been no change in their time budgets, while for the remaining three-quarters, some had more and some had less time for leisure activities. As a consequence it cannot be said that in 1991

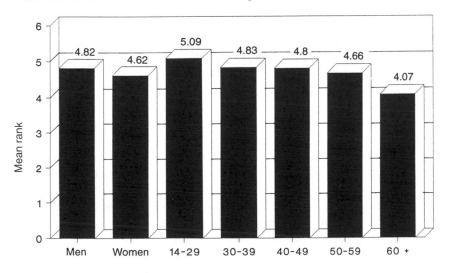

Fig. 5.2. The importance of leisure in relation to gender and age, 1991 (mean ranks).

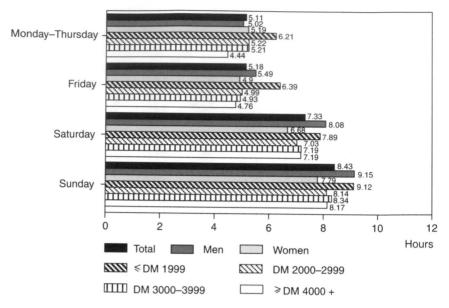

Fig. 5.3. Leisure time budgets during the week and on weekends in relation to gender and income in hours (averages).

people in Germany had more time for leisure activities than 5 years before. The differences in relation to gender, age, class and other socioeconomic factors are not significant. Having 'too much leisure time' is sometimes seen as a characteristic of modern lifestyles in general, but results of the ELS show that this is to some extent true for some groups, notably elderly people and people of lower socioeconomic status, and it is more true for men in general and young men in particular, than for women.

Leisure activities
Information on participation rates in some 81 leisure activities was gathered using a scale in which the respondents could indicate whether they had undertaken an activity often, sometimes, rarely or never. Rather than listing simple items, such as 'reading', 'being at home' or 'shopping', the list of leisure activities contained qualified items which described the milieu of the activity (for example 'reading a book' or 'being at home having fun with others' or 'shopping without time pressure'). For those respondents who indicated that they engaged in a leisure activity often or sometimes additional data, such as motivation, with whom and at what time the activity was done, were collected. In addition respondents were asked if they wanted to do the activity more often. The following analysis of individual leisure activities compares only the 'often' and 'never' responses in relation to each activity and compares

Table 5.1. Individual leisure activities in West and East Germany, 1991.

	Frequency of participation (%)			
	West Germany		East Germany	
	Often	Never	Often	Never
Read newspapers	64	5	68	2
Spend a pleasant time at home	49	6	53	6
Watch entertainment and sport on TV	48	8	47	6
Read magazines	31	7	40	5
Have friends and relatives to visit	29	8	29	4
Walk in the countryside	26	12	28	11
Do nothing, doze, rest	25	12	18	16
Watch political, cultural, scientific programmes on TV	25	17	32	10
Gardening	24	39	33	31
Play with children	19	40	29	27
Play with pets	18	60	20	60
Be active in sports	17	49	10	53
Eat/drink out	17	13	7	20
Concentrate on listening to music	16	28	20	28
Go shopping without time pressure	16	19	19	23
Read high quality books	15	34	14	32
Read for further education or go to evening classes	13	44	15	38
Stroll through town	12	15	17	14
Be active in a club/party/organization	12	62	10	71
Go to a swimming pool	11	33	10	38
Play cards etc.	10	31	8	37
Visit cultural sites/sightseeing	10	17	13	13
Do-it-yourself	9	44	13	47
Have a (second) job	9	68	3	75
Knit, sew	9	56	13	51
Go to church	9	48	6	70
Do social/voluntary work	9	49	13	39
Go to sport events	8	50	5	61
Make music/sing	6	72	6	75
Keep fit by jogging	6	61	3	73
Play/work on computer	5	81	4	85
Go to cinema	4	49	3	47
Go to theatre/concert	3	48	2	54
Drawing/painting/pottery	2	79	2	86

Source: EMNID Leisure Survey, 1991.

West and East Germany participation levels. Data relating to the most popular 34 activities are presented in Table 5.1, beginning with activities with the highest frequency of participation in West Germany.

The figures present a fairly familiar pattern, characteristic of Western European countries for many years. Home-based and social activities are the focus of leisure behaviour for the majority and are much more popular than the more participatory activities, such as sport, or more demanding activities, such as visits to cultural events (Veal, 1984).

It is clear that the differences between West and East are not as great as might have been expected. The former German Democratic Republic had been a developed industrial country with the highest living standard in the Eastern block. The *rapprochement* between the two parts of Germany on the field of leisure had started much earlier than 1990, as statistical analyses from the 1980s show (see Staatliche Zentralverwaltung für Statistik 1986).

Trends in participation cannot be discussed in detail here, but in general the patterns of participation shown by the 1991 survey have been fairly stable in Germany for many years, and can be seen as the typical leisure pattern of a western industrial society.

ACKNOWLEDGEMENTS

The authors acknowledge that the analysis was made possible by the support of Dr Henry Puhe of the EMNID Institute, Bielefeld.

REFERENCES

Staatliche Zentralverwaltung für Statistik (ed.) (1986) *Statistisches Taschenbuch der Deutschen Demokratischen Republik 1986*. Staatsverlag der Deutschen Demokratischen Republik, Berlin.

Veal, A.J. (1984) Leisure in England and Wales. *Leisure Studies* 3 (2), 221–230.

6 Great Britain

CHRIS GRATTON

INTRODUCTION

This chapter reviews statistics on British leisure participation as revealed in the General Household Survey (GHS), the major survey used for the collection of data on leisure participation in Britain. The GHS is an annual survey of around 20,000 adults aged 16 and over, which has been carried out by the government Office of Population Censuses and Surveys (OPCS) continuously since 1972. It is a wide-ranging survey covering topics of interest to government, such as housing, education and health. Questions on leisure activities have not been included every year but have appeared on a regular basis.

The first time leisure questions were included was 1973. Questions were asked on all aspects of leisure, including home-based leisure activities (gardening, TV, radio, listening to records, etc), various social activities, informal outdoor recreation trips, arts activities, entertainment, and watching and playing sport. Respondents were asked about their participation in the 4 weeks before interview, and interviewing for the GHS was conducted throughout the year. The results were made available on an annual or quarterly, or seasonal, basis. In this chapter annual average figures are generally used. The second time leisure questions were asked was 1977. However, substantial changes were made to the prompt card of activities shown to respondents, resulting in significant changes in participation rates in some activities; as a result 1977 has become the base year for comparative purposes. The methodology used in 1977 was repeated in 1980, 1983 and 1986 to give a consistent series of participation data over this period. By 1986, the list of leisure activities other than sports participation had become more restricted, mainly by limiting the range of other social activities and hobbies that were recorded.

—In 1987, questions on sports participation were again included in the GHS. However, several important methodological changes were introduced to the way data on sports participation were collected. In particular, instead of the conventional 'open-ended' leisure question used in previous surveys, a pre-coded sports and physical exercise section was introduced. A total of 32 categories of sport were prompted. The consequence of this second major change in the prompt card was that reported participation rates for some sports increased dramatically. Thus 1987 became a new base year. The new methodology was followed in the 1990 GHS with some minor modifications (most importantly an additional distinction was made between activities done mainly indoors and those done mainly outdoors) and these data form the basis for much of this chapter. However, as already indicated, the focus for the leisure section of the GHS in recent years has shifted towards sport and the 1990 data have little information on other leisure activities. The 1987 GHS report, however, had a special section covering arts and entertainment and there was a separate report on the sports participation data because of the change in methodology. Since the 1987 survey contained the most comprehensive information available on leisure participation, and since it is interesting because of the new methodology and can be compared with 1986, these data form the main source for this chapter.

The aim of this chapter is not only to report the basic data but also to examine trends in participation. To do this the second half of the chapter concentrates on sports participation and introduces a methodology for identifying the most important trends.

PARTICIPATION IN LEISURE ACTIVITIES

Table 6.1 gives the participation rates in a selection of home-based leisure activities that were included in the GHS on a consistent basis from 1977. For these social activities and hobbies there was little change in the overall pattern of participation over the 10-year period, although there was some growth (in particular, in listening to records, and DIY). The activities listed are all leisure activities with high participation rates. Nearly all adults watch television, visit or entertain friends, listen to the radio, and listen to records/tapes/ CDs at least once over a 4-week period.

Table 6.2, on the other hand, shows that participation rates in out-of-home leisure activities in the cultural area, including arts and entertainment and visiting galleries, museums, and historic buildings, are relatively low. Visiting the cinema is the most popular activity shown in Table 6.2 with an 11% participation rate, whereas only 1 or 2% of adults go to the opera, or

Table 6.1. Participation in selected home-based leisure activities: Great Britain, 1977–1987.

Activity	% participating in 4 weeks before interview,* persons aged 16 and over				
	1977	1980	1983	1986	1987
Watching television	97	98	98	98	99
Visiting/entertaining friends or relatives	91	91	91	94	95
Listening to radio	87	88	87	86	88
Listening to records/tapes	62	64	63	67	73
Reading books	54	57	56	59	60
Gardening	42	43	44	43	46
Do-it-yourself	35	37	36	39	43
Dressmaking/needlework/knitting	29	28	27	27	27
Sample size	23,000	22,600	19,000	19,200	19,500

Source: General Household Survey, OPCS (1989b).
* Annual average of four quarterly samples.

ballet, or classical or jazz concerts. The second column of Table 6.2 shows how frequently participants take part in these activities over a 4-week period. In all cases the frequency lies between one and two (i.e. once or twice a month).

Table 6.3 shows how participation in these cultural leisure activities varies with socioeconomic group (SEG). As with many leisure activities those

Table 6.2. Arts and entertainment and cultural visits, Great Britain, 1987.*

Activity	% participating in 4 weeks before interview, persons aged 16 and over	Average number of occasions of participation in 4 weeks, per participant
Arts and entertainment visits		
Films	11	1.6
Plays, pantomime, musicals	7	1.3
Ballet, modern dance	1	1.5
Opera or operetta	1	1.3
Classical music	2	1.5
Jazz, blues, soul, reggae	2	1.7
Other musical shows	7	1.9
Cultural visits		
Art galleries or museums	8	1.6
Stately homes, castles, other historical buildings	8	1.7
Sample	19,500	

Source: General Household Survey, OPCS (1989b)
* Annual average of four quarterly surveys.

in the professional group have the highest participation rates and those in the unskilled manual group have the lowest. Participation rates among the higher status socioeconomic groups are several times higher than those of the lower status groups, indicating considerable inequality.

The most comprehensive analysis of leisure participation data from the 1987 GHS has been the analysis of the sports participation data, with a separate report produced for these data (Matheson, 1990). Not only are there 12-monthly as well as 4-weekly participation rates for these sports data, but also, because of the change in survey methodology between 1986 and 1987 there was a substantial increase in reported participation for most activities, particularly outdoor activities, in 1987. The approach before 1987 seems to have led to some under-recording of sports participation and the 1987 data will provide a new benchmark for all future analysis of sports participation.

Table 6.4 reports the pattern of sports/physical recreation participation (4-weekly and 12-monthly) and the frequency (over 4 weeks) for 1987. Some 61% of adults took part in at least one sporting activity in 1987 in the 4 weeks before interview and 78% in the 12 months before interview. The activity with by far the largest participation rate was walking, with a 4-weekly participation rate of 38% and an annual participation rate of 60.1%. High participation rates were also recorded for snooker/billiards/pool (15% over 4 weeks; 23% over a year) and swimming (13% over 4 weeks; 35% over a year). However, for most of the activities listed in Table 6.4, the participation rate is less than 5% on a monthly basis and less than 10% on an annual basis. As in the arts, most sporting activities involve only a small percentage of the adult population.

Certain sports emerged for the first time as important participant sports in 1987. That is, the 1987 participation rates were substantially higher than those recorded previously. These include cycling, keep-fit/yoga, other running (including jogging), and weightlifting/weight training. The latter two were coded separately for the first time in 1987 and emerged immediately close to the top of the rankings of most popular sports. Keep fit/yoga is dominated by women with a 12% female participation rate over 4 weeks (compared with 5% in 1986) and 21% over a year. Cycling, which had a 4-week participation rate of only 2% in 1986, had a 4-week participation rate of 8% and an annual rate of 15% in 1987. These major changes do not reflect a sudden surge in demand for these sports: rather they illustrate the effects of the new survey methodology, in particular the interviewer prompting of these activities for the first time in 1987.

Another important feature of Table 6.4 is the relationship between 4-weekly and annual participation rates. The latter measure, obtained by asking respondents if they had participated in the previous 12 months, includes less frequent participants. It also takes some account of seasonality since the 4-weekly rate is an average of four quarterly, or seasonal, sets of data. If there is substantial seasonal variation, as there is in many outdoor

Table 6.3. Arts and entertainment and cultural visits by socioeconomic group, Great Britain, 1987.

	% participating in 4 weeks before interview,* persons aged 16 and over						
	Professional	Employers/ managers	Intermediate and junior non-manual	Skilled manual and self-employed	Semi-skilled manual and personal service	Unskilled manual	Total
Arts and entertainment visits							
Films	15	11	12	8	8	5	11
Plays, pantomimes or musicals	13	10	10	4	4	3	7
Ballet or modern dance	1	1	1	1	–	1	1
Opera, operetta	2	2	1	–	–	–	1
Classical music	6	3	3	1	1	1	2
Jazz, blues, soul, reggae	2	2	2	1	1	–	2
Other musical performances	9	7	8	7	6	5	7
Cultural visits							
Art galleries or museums	15	11	11	6	5	4	8
Stately homes, castles, historic buildings	14	11	10	6	5	4	8
Sample	705	2465	6012	4051	3830	1265	19,529

Source: General Household Survey, OPCS (1989b). Note: – indicates less than 0.5%; * annual average of four quarterly surveys. Respondents are classified according to their own occupation or that of the head of household.

Table 6.4. Sport and physical recreation participation, Great Britain, 1987.

	(a)	(b)	(c)	(d)
Walking (at least 2 miles)	37.9	60.1	1.6	8
Snooker/billiards/pool	15.1	22.9	1.5	6
Swimming: outdoor	3.5 }	34.6	–	4
Swimming: indoor	10.5 }			
Darts	8.8	15.4	1.8	6
Keep fit/yoga	8.6	14.3	1.7	9
Cycling	8.4	14.8	1.8	10
Athletics – track & field	0.5	2.0	4.0	5
Other running/jogging	5.2	10.5	2.0	7
Football	4.8	8.9	1.9	4
Weightlifting/training	4.5	8.2	1.8	8
Golf	3.9	9.2	2.4	4
Badminton	3.4	8.2	2.4	3
Squash	2.6	6.7	2.6	4
Table tennis	2.4	6.3	2.4	4
Fishing	1.9	5.8	3.1	3
Tennis	1.8	6.6	3.7	4
Ten-pin bowling/skittles	1.8	5.7	3.2	2
Lawn/carpet bowls	1.7	3.7	2.2	6
Cricket	1.2	4.2	3.5	3
Water sport (excl. sailing)	1.1	4.7	4.3	3
Horse riding	0.9	2.6	2.9	7
Self-defence (excl. boxing)	0.8	1.7	2.1	7
Ice skating	0.8	3.7	4.6	2
Basketball	0.6	1.7	2.8	3
Sailing	0.6	2.5	4.2	3
Motor sports	0.4	1.1	2.8	4
Rugby	0.4	1.1	2.8	5
Netball	0.4	1.4	3.5	3
Gymnastics	0.3	0.6	2.0	7
Boxing/wrestling	0.2	0.4	2.0	7
Hockey	0.2	0.3	1.5	5
Field sports	0.1	0.2	2.0	7
Climbing	0.1	0.2	2.0	4
Curling	–	0.2	–	9
Other	0.7	1.5	2.1	3
At least one activity (excl. walking)	44.7	61.9	1.4	–
At least one activity	60.7	77.6	1.3	–
Sample	19,529	19,529		

Source: General Household Survey, OPCS (1989b)
* Annual average of four quarterly surveys
(a) % participating in 4 weeks before interview,* persons aged 16 and over.
(b) % participating in 12 months before interview.
(c) ratio of annual to 4 weeks participation rate (b:a).
(d) number of occasions of participation per participant in 4 weeks.

sports, the average 4-weekly rate is reduced by the averaging procedure. Thus, for example, in the case of tennis, the 4-weekly participation rate in the third quarter (summer) is 4%, in the first quarter (winter) it is less than 0.5% and the average over four seasons is 1.8%. The 12-monthly, or annual, rate is 6.6% and therefore the ratio of annual to 4-weekly rates is 3.7 (6.6:1.8). Seasonal sports, as shown in Table 6.4, tend to have a high ratio of annual to 4-weekly participation rates.

The final column of Table 6.4 shows the frequency of participation over the 4 weeks before interview. Sports participants have much higher frequencies of participation than do visitors to arts events, as recorded in Table 6.2. The highest frequencies occur in cycling (ten occasions per 4-week period) and keep fit/yoga (nine occasions per 4-week period). Thus participation occurs in these sports on average once every 3 days. Since these figures represent averages over all participants, some participants will have considerably higher frequencies than this. This high frequency of participation across a large number of sports is an important characteristic of sports participation.

Table 6.5 shows the equivalent table to Table 6.4 but for 1990. In general, participation rates in individual sports in 1990 were very similar to those in 1987, with a small overall increase in participation. For instance, 65% of respondents in 1990 indicated that they had participated in at least one sport compared to 61% in 1987. Frequency of participation was also very similar in the 2 years.

THE CONVENTIONAL APPROACH TO THE ANALYSIS OF GHS SPORTS PARTICIPATION DATA

By the conventional approach is meant the standard official analysis that takes place each time a General Household Survey data set that includes leisure data becomes available. These analyses fall into two groupings. The first relates to the analysis carried out by the Office of Population Censuses and Surveys. In 1977, 1983 and 1986 leisure data from the GHS were included in the main GHS reports for those years and the tabulations were accompanied by a commentary on the major aspects of leisure behaviour revealed by the data. Both the tabulations and the commentary followed a similar pattern for each of the years, although in 1986 this commentary section was severely curtailed and covered less than two pages of the report compared with seven pages in 1983. In 1980, there was no chapter on the leisure data in the GHS report since the intention was for the Sports Council to publish an analysis of the leisure data separately. In fact this never happened and so there was never a published analysis of the 1980 GHS leisure data. In addition to these analyses carried out by the OPCS, the Sports Council commissioned research

Table 6.5. Sports, games and physical activities, Great Britain, 1990.

	(a)	(b)	(c)	(d)
Walking	40.7	65.3	8	39.9
Swimming: indoor	12.2	35.6	3	5.4
Swimming: outdoor	3.9	21.8	6	3.0
Swimming: total	14.8	42.4	4	8.2
Snooker/pool/billiards	13.6	21.7	5	9.2
Keep fit/yoga	11.6	18.8	9	12.9
Cycling	9.3	17.0	10	11.7
Darts	7.1	13.1	4	4.0
Golf	5.0	12.2	4	2.6
Running/jogging	5.0	9.5	6	4.0
Weight lifting/training	4.8	9.1	7	4.6
Soccer: outdoor	3.8	7.3	5	2.3
Soccer: indoor	1.8	4.4	3	0.8
Total soccer	4.6	8.5	5	3.1
Ten-pin bowling/skittles	3.8	11.4	2	1.0
Badminton	3.3	8.5	3	1.4
Squash	2.5	6.2	4	1.4
Lawn bowls	1.2	3.8	7	1.1
Carpet bowls	1.1	2.9	5	0.7
Total bowls	2.1	5.3	6	1.7
Tennis	2.0	7.4	4	1.0
Fishing	2.0	6.1	3	0.9
Table tennis	2.0	5.4	4	1.0
Water sports (excl. sailing)	1.2	5.0	3	0.5
Cricket	1.1	3.8	3	0.4
Horse riding	1.0	3.1	6	0.8
Field sports	0.8	2.0	3	0.3
Sailing	0.8	2.8	3	0.4
Self-defence	0.7	1.5	6	0.6
Ice skating	0.7	3.6	2	0.2
Basketball	0.6	1.7	3	0.2
Hockey	0.6	1.6	4	0.3
Climbing	0.5	2.5	3	0.2
Athletics – track and field	0.5	1.7	4	0.3
Rugby	0.5	1.1	5	0.3
Motor sports	0.4	1.2	3	0.2
Skiing	0.4	3.0	4	0.2
Netball	0.4	1.4	3	0.2
Gymnastics	0.2	0.5	5	0.1
At least one activity (excl. walking)	47.8	67.3		
At least one activity	64.5	81.9		
Sample	17,574	17,574		

Source: General Household Survey, 1990, OPCS (1992)
* Annual average of four quarterly surveys.
(a) % participating in 4 weeks before interview*, persons aged 16 and over.
(b) % participating in 12 months before interview.
(c) frequency of participation per participant in 4 weeks before interview.
(d) frequency of participation per adult per year.

into the GHS leisure data for each of the years the data became available. For 1977 and 1980, this was carried out by Veal (1979, 1984) and for 1983, it was carried out by The Henley Centre for Forecasting but not published. In these studies, the basic participation rate (the proportion of the sample recording participation in the activity over the 4-week reference period) has proved the most popular measure of sports participation. An additional indicator used here, the number of occasions of participation over the 4-week reference period, or the frequency of participation, is an important measure of the intensity of participation.

Not surprisingly the first interest of researchers in the analysis of the GHS leisure data was in the pattern of participation. The first questions that needed to be answered were: what are the most popular sports? Who are the people that take part (i.e. what are their characteristics – age, sex, SEG, etc.)? How frequently do they participate? How do these characteristics vary from sport to sport? Since the General Household Survey consists of cross-section (as opposed to time-series) data, it is particularly appropriate for answering such questions and the data have provided detailed answers to these questions, as the previous section illustrates.

Although cross-section analysis has been the major focus of interest for leisure researchers, there is now a sufficient time span of GHS sports data to allow the analysis of time trends. Some time trend analysis has already been carried out. The report on the 1986 GHS and Veal (1991) both analyse trends in participation from 1977 to 1986. Both concentrate on individual activities, although four aggregate participation groups are used, relating to overall participation in indoor and outdoor sporting activities. The most significant change between 1977 and 1986, according to these analyses, was in the participation of women in indoor sport (rising from 13% in 1977 to 21% in 1986), with other indicators showing a slow but steady increase in sports participation. Neither of these analyses picked up any major changes in the relationship between sports participation and other major demographic and socioeconomic variables. With this exception, the conventional analysis of the GHS data has indicated a fairly static pattern of association between sports participation and these variables.

For instance, the report on the 1983 GHS stated:

> The pattern of participation rates among the different age groups is similar to that found in the 1977 and 1980 surveys. Participation generally tends to be highest among those aged 16–19 and to decline with age.
>
> (OPCS, 1985)

Similarly, the report on the 1986 GHS stated:

> The pattern of participation rates among the different age groups in 1986 is similar to that found in the 1983 survey.
>
> (OPCS, 1989a)

—The conventional approach has been very successful in identifying the basic pattern of sports participation. We now know what the most popular participation sports are, how frequently participants take part, and how participation is related to major classification variables such as age, sex and socioeconomic group. There has, however, been less attention paid to trends in participation. —

There are two aspects of the conventional approach that tend to make it difficult to pick out major trends. The first relates to the activities included in the definition of sport. Camping/caravanning is included in the outdoor sport category of the GHS throughout the 1977 to 1986 period. There is little logic to this. This activity has a participation rate of 2% in the most popular (third) quarter, a significant amount for an activity with little or nothing to do with active sport. More seriously, two indoor activities dominate the indoor sport category for men: snooker, with a male participation rate of 17% in 1986, and darts, with a male participation rate of 9% in 1986 (but 15% in 1977). There is perhaps more of an argument for including these activities as sports than there is for camping/caravanning and they may be important activities for indicating the overall pattern of leisure participation. However, if the main interest is to indicate the trends in sports participation in order to assess the success of publicly financed policies aimed at encouraging people to take part in active sport, then it is much better to exclude these activities for two reasons.

First, because public subsidy to sport is based to some extent on the sports participation/health relationship, the definition of sport used in the statistical indicators used to measure the effectiveness of such subsidy policies should include those sports involving some positive aspect of health promotion. Similarly, excellence arguments relating to the promotion of international sporting success have little relevance in these activities. The second reason, however, is the most important. Any statistical indicator of sports participation will be influenced to a major extent by these two categories because of their size. Since one of them, snooker, grew considerably over the 1977 to 1986 period, but the other, darts, declined to almost an equal extent, statistical indicators containing them will show little or no change. This may be one of the reasons why there was apparently more change in the indoor sports participation of women than men, since darts is dominated by male participants.

The other major problem with the conventional analysis results from the initial interest in the data for the investigation of the pattern of participation. This initial interest led to the reporting of results for a large number of sports categories, some of which have very small sample sizes and hence are prone to large sampling errors. As a result only a few aggregate indicators of sports participation have been derived and all of these suffer from the problems raised above. There is a need for a series of sports participation indicators that represent the major features of the GHS data. These indicators should be capable of providing an adequate summary of the data in a relatively small set of statistics. By analysing the path of these indicators over time it would then become possible to analyse the major trends in the data. This is attempted in the next section.

TRENDS IN SPORTS PARTICIPATION: 1977–1986: AN ALTERNATIVE APPROACH

In this section an attempt is made at an alternative approach to the analysis of trends in GHS sports participation as revealed by the GHS data. The time period chosen for the trend analysis was 1977 to 1986 because this is the only period over which the GHS sports participation data are comparable.

The basis of the approach is to extend the use of aggregate participation groups from the four that have been used since 1983 to a wider range of participation indicators that together represent the major dimensions of sports participation behaviour. The four conventional groups were based on partici-pation in at least one of: (i) outdoor physical recreation activities including walking, (ii) outdoor physical recreation activities excluding walking, (iii) indoor physical recreation activities, (iv) all indoor and outdoor physical recreation activities. For this analysis some 14 new participation groupings were devised (Gratton and Tice, 1994). Here just six of the groups are used and, for clarity of exposition, walking is included as a sport in all six groups. Darts, billiards/snooker and camping/caravanning have been excluded for the reasons given above. The six participation groups that provide the structure of the analysis are defined in Table 6.6.

Table 6.7 reports the participation rates for these six groups, as well as the changes in participation for the 1977–1986 period. The first group, 'sport', representing overall sports participation, shows a rise in participation from 33 to 41% from 1977 to 1986. Thus participation in sport was increasing due to a higher proportion of the adult population taking part in at least one activity. However, examination of the other groupings shows that the major growth

Table 6.6. Sports participation groups.

Name	Description: respondents included in this group if they take part in:
1. Sport	at least one sport
2. Outdoor	at least one outdoor sport (including walking)
3. Indoor	at least one indoor sport
4. Outdoor only	outdoor sport only (including walking)
5. Indoor only	indoor sport only
6. Indoor & outdoor	both indoor and outdoor sport

Chris Gratton

Table 6.7. Sports participation rates, Great Britain, 1977–86.

Group	% participation in 4 weeks before interview*				% change 1977–86
	1977	1980	1983	1986	
1. Sport	33.3	36.4	38.2	40.7	22.2
2. Outdoor	27.6	29.7	30.7	31.4	13.8
3. Indoor	11.8	14.3	15.8	18.6	57.6
4. Outdoor only	21.5	22.1	22.5	22.1	2.8
5. Indoor only	5.8	6.7	7.5	9.3	60.3
6. Indoor & outdoor	6.1	7.6	8.2	9.3	52.5

* Annual average of four quarterly surveys.

was in indoor sports participation. Outdoor sports were not attracting a significantly greater number of participants in 1986 than in 1977. In particular, those people that only take part in outdoor sport represented the same share of the adult population in 1986 as they did in 1977.

The picture for indoor sports is remarkably different. 'Indoor' (i.e. participants in at least one indoor sport) has an overall growth rate of 58% between 1977 and 1986. The equivalent figure from the conventional GHS analysis was only 29%, indicating how the inclusion of darts and snooker seriously reduced the apparent growth rate in indoor sports participation. 'Indoor only' has a growth rate of 60%. There is therefore a clear difference in the growth of participation in indoor and outdoor activities. Gratton and Taylor (1991), in analysing the relationship between participation and the growth of public indoor swimming pools and sports centres over the 1970s and 1980s, concluded that it was the massive public investment in new indoor facilities that led to the large increase in participation in indoor sport in that period. The current analysis indicates that the increase was under-recorded by the conventional measure of indoor sports participation.

The definition of the participation groups in Table 6.6 has revealed clear patterns of participation and changes in participation that conventional analysis of the GHS data has not fully revealed. The major feature of the period 1977 to 1986 is the contrasting growth rates in participation between indoor and outdoor sports with the former far outstripping the latter. Such clear trends are much easier to identify by using a more meaningful definition of sport and by using a wider number of indicators covering different aspects of types of sports participation.

Table 6.8. Frequency of participation in sport, Great Britain, 1977–1986.

Activity group	Participants' frequency of participation in 4 weeks before interview*			
	1977	1980	1983	1986
1. Sport	8.2	8.9	9.7	9.3
2. Outdoor	9.0	9.7	10.8	10.5
3. Indoor	8.7	9.4	10.0	10.1
4. Outdoor only	7.9	8.5	9.5	8.7
5. Indoor only	4.5	4.9	5.2	5.3
6. Indoor & outdoor	12.7	13.3	14.5	14.9

* Annual average of four quarterly surveys.

FREQUENCY OF PARTICIPATION

Table 6.8 shows frequency of participation for the six activity groups. There is an increase in frequency of participation over the period 1977 to 1986 in all groups. The pattern of growth in frequency is, however, not as steady as for participation. Whereas the largest growth in participation in indoor sports occurred in the 1977–80 and 1983–86 periods, the largest growth in frequency of participation occurred in the 1980–83 period. In fact frequency actually decreased in the 1983–86 period in most groups.

In general, frequency of participation in outdoor sports is higher than that for indoor sports. The lowest frequency is in the 'indoor only' group. By far the highest frequency is for the 'indoor and outdoor' group. Although the participation rate for this category was only 9.3% in 1986 (Table 6.7), these participants took part in sport on average 15 times per 28 day period: that is, more often than every other day. This group therefore represents the most committed sports participants.

CHARACTERISTICS OF PARTICIPANTS

In this section participation in the six participation groups is analysed with regard to three variables: age, gender and retirement.

Age

Table 6.9 records the average age of participants for the different participation groups for the four surveys. The pattern is remarkably consistent from year to year and across groups, with a gradual increase in the average age of participants. Outdoor sports participants are, on average, between 7 and 10

Table 6.9. Age of sports participants, Great Britain, 1977–86.

Activity group	Average age in years			
	1977	1980	1983	1986
1. Sport	38.2	38.3	38.9	39.0
2. Outdoor	39.3	39.5	39.9	40.4
3. Indoor	32.2	32.0	33.4	33.9
4. Outdoor only	41.5	42.4	42.7	43.3
5. Indoor only	33.2	33.0	34.5	34.2
6. Indoor & outdoor	31.3	31.2	32.3	33.7
Average (whole sample)	45.4	45.3	45.7	45.5

years older than indoor sports participants. The consistency in the way average age of participation has changed over time is quite remarkable. All groups show a steady rise over the period.

This provides some evidence in support of Rodgers' (1977) hypotheses concerning the relationship between age and sports participation. Rather than accepting the conventional view that people 'drop out' of active participation as they get older, he suggested that many people in the older age categories were never participants, even when young. He introduced the concept of 'sports literacy', the extent to which sections of the population are exposed and conditioned to active recreation. The older generation, he suggested, reveals a much higher proportion of illiteracy than the younger generation. Table 6.9 suggests that between 1977 and 1986, as a more 'sports literate' generation got older, they continued to be involved in sport, so that many older people were regular sports participants in 1986 than 10 years earlier. If this interpretation is correct the implication is that the average age of participants will continue to rise gradually over the foreseeable future.

Gender

Table 6.10 shows the percentage of female participants in each of the six groups. In 1977 just over 42% of all participants were women; by 1986 this had risen to nearly 46%. The table, however, shows that this overall rise conceals a varied pattern of change among the different groups. In particular there was little change in the proportion of female participants in the outdoor sport groups.

In contrast to the outdoor groups, the proportion of women in the indoor sports participation groups increased dramatically so that, by 1986 women outnumbered men in indoor activities. The proportion of females in the most committed group of sports participants, 'indoor and outdoor', rose from just

Table 6.10. Percentage of female sports participants, Great Britain, 1977–1986.

Activity group	% of participants who are female				% change 1977–86
	1977	1980	1883	1986	
1. Sport	42.4	44.5	45./	45.6	7.6
2. Outdoor	40.7	42.5	42.6	41.1	1.0
3. Indoor	42.4	45.8	50.1	51.4	21.2
4. Outdoor only	42.4	43.6	42.6	40.6	-4.3
5. Indoor only	50.5	53.5	58.6	60.5	19.8
6. Indoor & outdoor	34.7	39.1	42.4	42.3	21.9
Whole sample	53.4	53.6	54.1	53.7	

over 34% of the total in 1977 to over 42% by 1986, the biggest proportionate rise of any of the groups.

The 'sports literacy' argument discussed in the last section in relation to age could also be part of the explanation for the rising level of female participation in sport. Historically women have been less sports literate than men. However, this is less and less true in the latter quarter of the twentieth century. Women increasingly have similar exposure and opportunities for sport to men. This is starting to be expressed in the statistics, but more for indoor sport than outdoor. There is interaction here, however, between age and gender. Younger women are showing trends in participation similar to those of men, but the difference in sports literacy between older men and women results in continuing disparities between male and female participation rates for these age groups.

Retirement

Age is also important. Table 6.11 which shows the percentage of participants who are retired. The table shows a large rise in the percentage of retired participants in all the participation groups, with the biggest percentage increases occurring in indoor activities and in the category of sports that represents the most committed sports people, 'indoor and outdoor'. Although the proportion of retired people in the population is increasing, the rate of increase of their numbers among sports participation groups is faster. Absolute levels of percentages of retired in each of the participation groups in Table 6.11 are, nevertheless, quite low except for the outdoor categories. That is, although participation by the retired is increasing most rapidly in the indoor sports categories, it is outdoor sport, and particularly walking, that attracts the highest numbers of retired sports participants.

Table 6.11. Percentage of retired sports participants, Great Britain, 1977–1986.

Activity group	% of participants who are retired				Change 1977–86
	1977	1980	1983	1986	
1. Sport	6.5	7.5	9.2	10.1	55.4
2. Outdoor	7.4	8.5	10.2	12.0	62.2
3. Indoor	1.9	2.3	4.0	4.5	136.8
4. Outdoor only	9.1	10.9	12.8	14.9	63.7
5. Indoor only	2.3	3.2	4.9	3.9	69.6
6. Indoor & outdoor	1.6	1.5	3.3	5.1	218.8
Whole sample	11.6	12.8	16.0	17.5	50.9

INADEQUACIES OF GHS DATA FOR RESEARCH INTO LEISURE PARTICIPATION

This summary of the Gratton and Tice (1994) research into trends in sports participation over the period 1977–1987 is intended to be indicative of progress that can be made through secondary analysis of existing data on leisure participation. Further investigation using the GHS was also undertaken, involving the estimation of sports participation demand models using multivariate techniques, but there is no space to record these results here.

This research revealed some of the inadequacies of the GHS data for serious research into leisure participation. The major inadequacies relate to the following.

1. Inability to relate sports participation to other areas of leisure lifestyle, in particular tourism and the arts. The GHS data has increasingly focused on sports participation so that it now does not provide a comprehensive database for research into leisure lifestyles.

2. Lack of variables to facilitate the creation of 'leisure types'. Despite developments in the tourism literature in terms of psychographic segmentation of leisure consumers, the GHS continues to adopt a socioeconomic/demographic approach which restricts the level of analysis possible.

3. Lack of expenditure data (on sports equipment, clothing, shoes, etc). It is possible to obtain only a limited picture of consumer behaviour because participation rates cannot be related to other indicators such as consumer expenditure.

4. Lack of time expenditure. Although frequency of participation is one measure of the intensity of participation, it is also important to know how long participants spend on average on each occasion of participation.

5. Lack of data on supply of facilities facing respondents (e.g. distance to nearest sports centre). This is probably the biggest single problem with the

GHS data. Participation rates for individual sports are not an indicator of demand for sport: they are indicators of the outcome of the interaction between demand and supply. Without information on the supply of facilities faced by the consumer it is not possible to separate out the different influences of demand and supply. This prevents any serious estimation of demand models using these data.

These problems relate specifically to the GHS and its use for research into participation and lifestyle. In fact *no* data set currently exists in Britain which is adequate for the testing of the theories of lifestyle and participation that have been developed over the last 20 years. In the 1960s, two surveys were carried out that were specifically targeted at research into participation and lifestyle, the *Pilot National Recreation Survey* (British Travel Association/ University of Keele, 1967) and *Planning for Leisure* (Sillitoe, 1969). At that time there was little information available and few attempts had been made at developing theories. In the 1990s, the GHS is still collecting very similar types of information to those earlier surveys even though leisure researchers have moved on considerably in theory formulation.

The most detailed and comprehensive study so far conducted into participation and lifestyle has been the Roberts and Brodie (1992) study into inner city sport. This study included new data that expanded our knowledge and understanding of sports participation, most importantly by examining past patterns of participation (or 'sports careers'). This work provided important insights into why we observed the changing relationship of sports participation with age discussed above. Such work should provide the starting point for developing a new leisure participation survey at national level that will rectify the inadequacies of current national data sets and allow leisure researchers to develop more sophisticated analyses of leisure participation and lifestyle.

NOTE

This chapter includes some of the results of a research programme, funded by the Health Promotion Research Trust, and carried out at Manchester Metropolitan University from 1988 to 1991. The full results are available in Gratton and Tice (1994).

REFERENCES

British Travel Association/University of Keele (1967) *The Pilot National Recreation Survey.* BTA/Univ. of Keele, London and Keele.
Gratton, C. and Taylor, P. (1991) *Government and the Economics of Sport.* Longman, Harlow.

Gratton, C. and Tice, A. (1994) Trends in sports participation 1977–1987. *Leisure Studies* 13 (1) 49–66.

Matheson, J. (1990) *General Household Survey 1987: Participation in Sport*. HMSO, London.

OPCS (Office of Population Censuses and Surveys) (1985) *General Household Survey 1983*. HMSO, London.

OPCS (Office of Population Censuses and Surveys) (1989a) *General Household Survey 1986*. HMSO, London.

OPCS (Office of Population Censuses and Surveys) (1989b) *General Household Survey 1987*. HMSO, London.

OPCS (Office of Population Censuses and Surveys) (1992) *General Household Survey 1990*. HMSO, London.

Roberts, K. and Brodie, D. (1992) *Inner-city Sport: Who Plays and What are the Benefits?* Giordano Bruno, Vooorthuizen, Holland.

Rodgers, B. (1977) *Rationalising Sports Policies: Sport in its Social Context*. Council of Europe, Strasbourg.

Sillitoe, K.K. (1969) *Planning for Leisure*. Government Social Survey, HMSO, London.

Veal, A.J. (1979) *Sport and Recreation in England and Wales: An Analysis of Adult Participation Patterns in 1977*. Research Memorandum 74. Centre for Urban and Regional Studies, University of Birmingham.

Veal, A.J. (1984) Leisure in England and Wales. *Leisure Studies* 3 (2), 221–230.

Veal, A.J. (1991) National Leisure surveys: the British experience 1977–86. In: Veal, A.J., *et al.* (eds) *Leisure and Tourism: Social and Environmental Change*. University of Technology, Sydney/World Leisure and Recreation Association, Sydney, pp. 414–417.

7 Hong Kong

ATARA SIVAN AND BOB ROBERTSON

INTRODUCTION

Unlike many western industrial nations, where numerous studies of leisure behaviour have been undertaken over recent years, few such studies have been conducted in Hong Kong. The few completed during the last three decades focused primarily on youth, generally seeking data to plan for the provision of community services for young people of the Hong Kong Territory. Most were small scale surveys undertaken by government and voluntary agencies to examine leisure activities undertaken in specific districts of Hong Kong or by specific age groups. The findings of these studies of leisure and recreation generally appeared in government reports on the social needs of particular districts. Seven of these early studies are briefly summarized below.

EARLY SURVEYS

In 1965, the Hong Kong Council of Social Service sought the views of young people in the age cohorts 15–19 and 20–29 on leisure (Hong Kong Council of Social Service, 1965). The majority of respondents regarded leisure activities as important and were frequent visitors at community centres. It was found that more respondents in the 15–19 age group than in the 20–29 group participated in physical recreation activities. For both groups, most leisure activities were undertaken together with friends and away from the family.

© 1996 CAB INTERNATIONAL. *World Leisure Participation* (eds G. Cushman, A.J. Veal and J. Zuzanek)

A 1967 study of participation in leisure activities by youth labourers in Hong Kong (reported in Hong Kong Baptist College, 1970) was undertaken using a sample of 666 young workers from 23 youth centres. Respondents ranged in age from 14 to 21 years and were mostly semi-skilled or unskilled labourers working an average of 10 hours per day. Respondents indicated a preference for participation in outdoor leisure activities. However, many actually spent most of their leisure time at home. The results of this research indicated that social outings and going to the movies were the most popular leisure activities away from home, while reading, listening to the radio and watching television were the most common indoor activities.

A subsequent study of the leisure activities of labourers between the ages of 14 and 21 was undertaken by the Hong Kong Baptist College in 1970. Some 390 young workers in two factories were selected, using cluster sampling techniques, and a structured interview, using 'show cards', was used to discover current and preferred leisure activities. Data on socioeconomic characteristics and information about respondents' residential environment were also collected. This study revealed that the most frequent leisure activities were watching television at home, studying at evening schools and playing ball-games outdoors with friends. On days free from work and on holidays, respondents most frequently participated in activities such as picnicking or social outings, going to movies, window shopping and undefined physical recreation activities. The most desired activities were, in order of preference: touring, camping, attending evening schools, sewing, picnicking and social outings. No statistically significant relationships were found between patterns of activity and respondents' age, gender, religious affiliation, occupation, skill or income (in some cases only pocket-money). In addition, no significant relationships were found between type of leisure activity and type of home residence, number of family members or birth order. Nor was any relationship found between availability of community recreational facilities and respondents' activity patterns.

A survey of children and youth in Kowloon City was undertaken in 1970 to examine leisure activities and attitudes towards youth service agencies (Kowloon City District, 1975). Random sampling techniques were used to select 1720 respondents from schools, youth service agencies and factories. Data were collected through an interviewer administered structured questionnaire which elicited information on basic socioeconomic and demographic characteristics, participation in leisure activities and knowledge of, attitudes towards, and participation in, summer activity programmes organized by schools, youth service agencies and factories. Results showed that the leisure activities most frequently undertaken were watching television, reading and ball-games. Watching television was the most frequent activity for both sexes. Among males, the second most popular activity was ball-games and reading rated third, whereas, among females, reading was the second most popular activity, followed by ball-games. Some 65% of respondents indicated that they would prefer to undertake outdoor activities, sports and ball-games, but did not have the opportunity. Some 84% of respondents had participated in school-based leisure activities.

A study of the leisure activities of youth in the Southern District of Hong Kong was carried out by the Hong Kong Southern District Board Social Service Committee in 1985. A random sample of young people aged between 12 and 24 received a questionnaire; 262 were returned – a 25% response rate. The results indicated that the ten most frequently pursued leisure activities were, in descending order of frequency of participation: ball-games, swimming, watching television, cycling, listening to records, watching movies, listening to the radio, camping, music and computers. Frequency of participation in all of these activities was high. Among activities in which respondents wished to participate but did not, were: camping, overseas tours, ball-games, windsurfing, watching movies, extra-mural courses, rowing, track and field sports, archery and ice-skating.

Research on the leisure behaviour and sociodemographic characteristics of adolescents in Hong Kong was undertaken in 1984 by Ng (Ng, 1984). This work was based on a self-administered questionnaire survey of 1400 secondary school students and revealed the relative popularity of various leisure activities. This was a more sophisticated study than those described above, and specifically sought to examine differences in leisure patterns according to gender, age and family and socioeconomic status. Data were collected using a pre-coded list of activities. Respondents were asked to indicate whether they had participated in any of the listed activities during the last month and to indicate the three types of leisure activity on which they spent the most time during the previous month.

The results, as shown in Table 7.1, again showed the domination of television watching. The study also revealed a clear gender differential in the most popular leisure activities. More females than males watched television and listened to the radio or records. Females were also more involved in social activities and household skills. Males were more involved in arts and crafts and in sports activities. Differences were also found in the frequency of participation of respondents in different age groups. Interestingly, there was a decline in watching television with increasing age, while the use of other media increased. As might be expected, participation rates in physical activities were higher among younger respondents than among those aged 16 and above.

A subsequent study of the leisure behaviour and life satisfaction of young people in the eastern District of Hong Kong was undertaken by Ng in 1987–88 (Ng and Man, 1988). The sample comprised 250 young people aged between 12 and 25 years old, selected by household survey, and an interviewer administered questionnaire was used to collect the data. In this study, a distinction was made between weekday and weekend activity and between summer and winter activities. The results indicate that television watching

Table 7.1. Leisure activities of young people, Hong Kong, 1984 (Sample, n = 1310).

	% participation in previous month
Watch television	31.5
Listen to radio/records	8.1
Movies	3.8
Read books	9.1
Read newspapers, magazines	2.8
Household skills	2.0
Art and craft	2.4
Performing arts	1.8
Social activities	9.2
Ball games	13.2
Outings and camping	7.6
Other sports	3.4
Other activities	5.1

Source: Ng, 1984.

was the most popular activity among the younger respondents, aged 12–17. Older respondents, aged 18–25, participated more in watching films and were inclined to use their leisure for resting and sleeping. For both age groups, swimming and other sports were common activities in summer when they participated more in out-of-home activities than during winter. The older respondents participated more in out-of-home activities than did the younger respondents.

1993 SURVEY

In 1993 the Hong Kong Sports Development Board commissioned the first comprehensive survey of recreation participation in Hong Kong, covering all age groups (Sivan and Robertson, 1993). This section provides a brief overview of the survey results.

The study was conducted in April and May, 1993, and collected data on:

- the range of leisure, recreation and sports activities participated in during the previous month;
- frequency of participation in these activities;
- place and organizing body for the activities;
- sources of information used;
- level of satisfaction with facilities used;
- activities respondents wanted to participate in but had not;
- reasons for non-participation in preferred activities.

Table 7.2. Leisure and sport activities by gender, Hong Kong, spring 1993.

	% participating in month before interview		
	Males ($n = 1203$)	Females ($n = 1714$)	Total ($n = 2917$)
Leisure Activities			
Watching TV	56.7	57.5	57.2
Shopping	30.8	49.2	41.7
Going to public library	38.0	42.7	40.6
Eating out	32.3	34.1	33.2
Listening to radio/records	27.5	31.6	29.9
Reading newspapers/magazines	26.3	28.9	27.8
Going to cinema	28.1	27.3	27.7
Playing computer/electronic games	38.7	16.8	25.8
Going to tea-houses	21.0	21.8	21.5
Picnic/barbecue	17.8	20.1	19.1
Karaoke	15.9	20.4	18.6
Reading books	16.5	17.4	16.9
Walking in park	17.0	15.5	16.1
Playing cards/Mahjong	18.5	12.3	14.8
Relaxing/doing nothing	16.0	12.5	13.9
Household activities	9.7	11.6	10.8
Seeing exhibitions/museums	8.3	8.5	8.4
Playing musical instruments	4.9	9.6	7.6
Taking courses	5.2	6.2	5.9
Religious activities	4.7	6.7	5.8
Arts/crafts	3.6	7.1	5.6
Watching horse racing	10.0	0.9	4.6
Popular music concerts	4.2	3.9	4.0
Voluntary work	3.0	3.5	3.3
Classical music concerts	3.2	1.6	2.3
Kite flying	2.8	1.3	2.0
Watching drama/dance	1.7	2.3	2.0
Meetings of cubs/scouts	2.1	1.6	1.8
Opera	1.6	1.5	1.6
Discotheque	1.1	1.1	1.1
Sauna/health massage	1.1	0.7	0.9
Visiting night clubs	0.6	0.3	0.4
Sports			
Badminton	43.1	57.7	51.6
Cycling	40.6	42.3	41.5
Basketball	43.1	32.3	36.8
Swimming	34.0	30.4	31.8
Table tennis	42.7	22.5	30.7
Jogging	21.4	28.1	25.4
Playground games	15.4	25.7	21.4
Soccer	42.7	2.9	19.2
Volleyball	12.5	23.5	18.8

Table 7.2. *contd*

	% participating in month before interview		
	Males (*n* = 1203)	Females (*n* = 1714)	Total (*n* = 2917)
Hiking	15.6	16.6	16.2
Tennis	9.6	8.8	9.1
Aerobics	4.3	12.4	9.0
Squash	7.4	9.8	8.9
Athletics	10.0	7.9	8.7
Ice skating	4.5	9.3	7.3
Bowling	7.4	6.7	7.0
Roller skating	2.7	8.9	6.4
Snooker	11.1	2.8	6.2
Dancing	0.9	9.2	5.7
Mountaineering	5.6	4.7	5.1
Boating	4.7	4.1	4.4
Yoga	3.0	5.1	4.2
Handball	4.7	2.3	3.2
Judo	4.9	1.5	2.9
Rifle	4.0	1.3	2.4
Diving	1.7	0.9	1.3
Boxing	2.0	0.6	1.2
Horse riding	1.9	0.6	1.2
Softball	0.7	1.6	1.2
Cricket	1.2	1.0	1.1
Lawn bowls	0.7	0.6	0.6
Rugby	0.7	0.5	0.6
Netball	0.2	0.4	0.3
Fencing	0.5	0.2	0.3
Water skiing	0.2	0.2	0.2

Source: Sivan and Robertson, 1993.

Questionnaires were distributed via educational institutions in the 19 administrative districts of Hong Kong. Selected students were given two questionnaires and asked to complete one themselves and to administer one to an adult family member. Some 3800 questionnaires were distributed and 2941 returned, a response rate of 78%. Because of the sampling method used, the resultant sample is biased towards young people, with 64% of the sample aged under 20.

Table 7.2 shows the level of participation in 35 sporting activities and 32 other leisure activities, by gender. The results confirm the findings of earlier surveys in terms of the most popular activities, but cover a much wider range of activities and age groups. While watching television is the most popular

Table 7.3. Participation in selected leisure and sport activities by age, Hong Kong, spring 1993.

Age groups.	% participating in month before interview								
	6–11	12–16	17–18	19–24	25–34	35–44	45–54	55–64	65+
	(n = 462)	(n = 1015)	(n = 178)	(n = 231)	(n = 164)	(n = 642)	(n = 187)	(n = 30)	(n = 11)
Leisure activities (top 12)									
Watching TV	62.1	58.4	65.7	41.6	59.8	55.0	55.6	53.3	72.7
Shopping	43.5	42.5	31.5	29.0	44.5	47.4	37.4	30.0	54.5
Going to public library	60.4	47.9	48.9	48.5	23.8	22.4	19.3	10.0	9.1
Eating out	29.2	27.1	23.6	42.4	40.2	42.8	36.9	23.3	27.3
Listening to radio/ records	21.2	38.9	51.7	37.2	20.1	19.9	17.6	26.7	36.4
Reading newspapers/ magazines	14.9	19.3	27.5	29.9	36.6	43.3	41.7	43.3	27.3
Going to cinema	18.0	34.0	35.4	53.7	34.1	17.9	10.2	6.7	0.0
Playing electronic/ computer games	41.3	40.8	29.8	16.9	8.5	5.9	2.7	3.3	0.0
Going to teahouse	18.8	14.4	9.0	17.7	26.8	32.4	36.4	46.7	45.5
Picnic/barbecue	22.1	19.8	16.9	18.2	14.6	19.0	17.1	10.0	9.1
Karaoke	15.2	21.0	23.6	30.3	25.6	12.9	9.6	6.7	18.2
Walking in park	19.9	11.0	6.7	7.8	14.6	23.1	27.3	23.3	36.4
Sports (top ten)									
Badminton	76.0	58.3	55.6	42.0	46.3	38.6	21.4	10.0	18.2
Cycling	58.2	47.2	38.8	33.8	34.1	33.8	20.3	16.7	9.1
Basketball	52.6	58.0	33.7	22.9	13.4	12.1	15.0	6.7	–
Swimming	42.2	32.6	21.3	23.4	31.1	33.0	25.7	10.0	9.1
Table tennis	47.4	39.2	26.4	14.7	17.7	21.2	17.6	6.7	9.1
Jogging	23.4	22.2	18.0	13.4	28.7	36.0	28.9	36.7	27.3
Playground games	33.3	22.0	10.1	11.3	19.5	22.0	12.8	20.0	18.2
Soccer	31.8	24.9	20.8	16.0	9.1	9.2	9.1	–	–
Volleyball	21.6	33.7	30.9	9.5	3.7	2.8	2.1	–	18.2
Hiking	8.9	11.1	19.7	24.2	23.8	22.7	19.7	20.0	9.1

Source: Sivan and Robertson, 1993.

activity, at 57%, the level of participation is much lower than in the western nations, despite a high level of television set ownership in Hong Kong. The inclusion of such activities as shopping, eating out, going to tea-houses and karaoke among the most popular activities, confirms the informal and social nature of leisure in Hong Kong, as elsewhere.

The high representation of young people in the sample results in artificially high levels of participation in sporting activities. However, as shown in Table 7.3, participation levels in many sporting activities are substantial among older age groups.

Table 7.4. Participation in selected leisure and sport activities by income, Hong Kong, spring 1993.

	None (n = 43)	<2 (n = 57)	Monthly Household Income, $HK '000s 2-3.9 (n = 150)	4-5.9 (n = 286)	6-7.9 (n = 435)	8-9.9 (n = 708)	10-14.9 (n = 384)	15-19.9 (n = 231)	20-24.9 (n = 125)	25-29.9 (n = 90)	30-34.9 (n = 71)	35+ (n = 141)
Sports (top 10)												
Badminton	62.8	48.2	59.6	58.4	54.2	51.3	47.3	47.0	46.4	46.7	54.9	4.0
Cycling	39.5	35.7	46.4	43.0	41.2	41.7	40.3	41.8	37.6	27.8	42.3	35.5
Basketball	41.9	30.4	45.7	37.8	39.6	36.1	35.1	31.9	31.2	31.1	40.8	32.6
Swimming	16.3	25.0	32.5	30.8	35.2	32.3	30.4	28.9	33.6	36.7	36.6	27.7
Table tennis	48.8	28.6	34.4	28.7	31.7	29.5	23.4	34.1	28.8	31.1	50.7	33.3
Jogging	25.2	25.0	25.2	27.6	25.7	25.1	28.3	21.6	21.6	27.8	28.2	21.3
Playground games	20.9	25.0	22.5	20.6	23.1	21.6	21.6	20.7	17.6	14.4	15.5	17.7
Soccer	41.9	12.5	27.2	17.5	20.1	18.8	18.4	15.9	16.8	16.7	18.3	22.1
Volleyball	20.9	17.9	28.5	22.0	21.3	19.2	15.3	13.4	13.6	23.3	14.1	7.8
Hiking	14.0	8.9	13.2	15.0	15.0	15.7	18.7	21.1	19.2	12.2	22.5	21.3
Leisure activities (top 12)												
Watching TV	46.5	55.4	60.3	63.6	57.2	57.1	56.9	57.3	56.0	53.3	49.3	49.6
Shopping	39.5	19.6	41.7	44.8	44.0	40.2	43.6	35.8	36.0	42.2	36.6	44.7
Going to public library	48.8	42.9	44.4	39.2	41.0	41.5	38.2	42.7	34.4	35.6	47.9	35.5
Eating out	20.9	28.6	26.5	31.1	30.1	32.0	35.8	40.9	32.0	43.3	47.9	39.0
Listening to radio/records	27.9	30.4	31.1	32.9	32.6	30.2	26.8	27.6	32.0	27.8	29.6	23.4
Reading newspapers/magazines	11.6	26.8	27.2	27.3	31.5	29.8	25.2	28.4	28.8	37.8	29.6	25.5
Going to cinema	25.6	30.4	31.1	22.0	25.7	28.9	30.1	26.3	29.6	30.0	33.8	22.0
Playing electronic/computer games	18.5	26.8	27.2	22.4	24.8	24.7	27.0	25.4	23.2	25.6	16.9	32.6
Going to teahouse	25.6	21.4	22.5	23.8	21.8	20.0	22.9	25.4	19.2	26.7	16.9	17.0
Picnic/barbecue	30.2	16.1	25.2	16.4	18.1	18.8	17.9	21.1	18.4	20.0	22.5	17.0
Karaoke	14.0	12.5	17.2	14.3	16.9	18.1	24.2	19.0	23.2	20.0	21.1	19.1
Walking in park	20.9	17.9	19.9	18.5	19.2	15.5	15.6	12.5	16.0	7.8	14.1	13.5

Values are percentage participating in month before interview.
Source: Sivan and Robertson, 1993.

Table 7.4 shows participation rates related to household income. Among the sports there is very little variation in participation rates across the income range, with two exceptions. Playground games show an inverse relationship between income and participation, which might be expected, since lower income groups tend to have less space in their homes and make more use of playgrounds. Hiking shows a positive relationship between participation and income, again, as might be expected, since hiking requires resources to travel outside of the urban area.

CONCLUSIONS

The leisure surveys so far conducted in Hong Kong reveal many similarities in participation rates to those revealed by surveys in western nations, but also some striking differences. The role of the family in leisure decision-making, the importance of shopping and eating out, and the use of tea houses and karaoke facilities are areas where the behaviour of Hong Kong residents appears to reflect eastern cultural imperatives, albeit overlain with strong western influences. Planners and managers of leisure facilites in Hong Kong, both public and commercial, should be sensitive to these apparent differences in making decisions about the provision of new leisure opportunities.

The authors, with the support of the Hong Kong Sports Development Board, are in the process of completing further survey work to verify earlier results and to further examine some of the unique features of leisure in Hong Kong.

REFERENCES

Hong Kong Baptist College (1970) *A Study of Leisure Activities of Youth Labourers in Hong Kong*. Hong Kong: Department of Sociology and Social Work, HKBC.
Hong Kong Council of Social Service (1965) *Chai Wan Social Needs Study*. Hong Kong: HKCSS.
Hong Kong Southern District Board Social Service Committee (1985) *Survey Report: How Youth Spend their Leisure Time*. Hong Kong: Hong Kong Southern District Board.
Kowloon City District (1975) *A Study of Children and Youth*. Hong Kong: KCD.
Ng, P.P. (1984) *Socio-demographic Patterns of Leisure Behaviour of Adolescents in Hong Kong*. Institute of Social Studies, The Chinese University of Hong Kong, Shatin, Hong Kong.
Ng, P.P. and Man, P.J. (1988) *Leisure Behaviour and Life Satisfaction of Youth in Eastern District, Hong Kong*. Institute of Social Studies, Chinese University of Hong Kong.
Sivan, A. and Robertson, R.W. (1993) *The Use and Demand for Recreational and Sports Facilities and Services in Hong Kong: Phase One Report*. Report to the Hong Kong Sports Development Board, Hong Kong Polytechnic, Kowloon.

8 Israel

HILLEL RUSKIN AND ATARA SIVAN

NATIONAL LEISURE PARTICIPATION SURVEYS IN ISRAEL

National leisure participation surveys in Israel have been conducted primarily by a group of researchers from the Israel Institute of Applied Social Research, led by Professor Elihu Katz. Katz and his colleagues conducted two major studies, in 1970 and in 1990, both commissioned by the Ministry of Education and Culture.

The first study was conducted in 1970 (Katz and Gurevitch, 1976) and was based on nearly 4000 personal interviews with a representative sample of Jewish adults aged 18 years and over. The respondents were selected at random from 56 different localities, which were themselves chosen from a stratified listing of all localities, ranging from the largest cities to the smallest villages, including different types of settlements. All cities with a population of over 20,000 were represented in the study. Among the smaller localities, the researchers preferred to cluster more respondents in fewer settlements rather than fewer respondents in a greater number of settlements, so that the cultural life of these settlements could be accurately represented.

Three kinds of data were obtained from the study: the leisure behaviour of the population over the 24 hours before the interview, including a time budget; the attitudes of the population towards variables such as the adequacy of leisure time and of facilities available for leisure, or the relative preference for different leisure programmes; and the perceived functions of various leisure and cultural activities and institutions, ranging from the mass media to books on different subjects and traditional and secular holidays.

A pioneering aspect of this study, as compared with other studies of leisure and culture which have been conducted both in Israel and elsewhere, is the attempt to integrate data on patterns of consumption of culture with data on the supply of cultural services. In analysing the data, the researchers chose to give primary emphasis to the variables that seemed most important to them: education and socioeconomic status, age, ethnicity, city size, religiosity and gender. The object of the study was also to contribute towards policy making in the fields of leisure, culture and communication. The study's findings did indeed have an effect on government policy makers with regard to the provision of facilities and services and in the determination of priorities.

A second national study was conducted in 1990 by the same group of researchers (Katz *et al.*, 1992). Demographically, the study population had grown from 2.5 million in 1970 to 3.5 million in 1990; the mean level of education had risen from 8.8 to 11.6 years of study; and the percentage of people with a higher education (post-high school) had doubled from 15% to 31%. During the period, the standard of living rose very significantly, and Israel became a typical western-style consumer society. Time and energy saving devices resulted in increased free time. More discretionary income and more discretionary time affected lifestyles quite significantly. Television, which was introduced into Israel in 1969, exerted a major effect on patterns of leisure behaviour. Also, Israeli society (50% of the employed population) moved from a $5\frac{1}{2}$-day work week to a five-day work week. All of these factors, and the effects of three wars which were fought in this region between 1970 and 1990, affected patterns of leisure behaviour in Israel.

The 1990 study was conducted in three stages. In the first stage, 'cultural indicators' were determined for an annual, on-going survey of leisure behaviour. The survey investigated patterns of cultural consumption, factors affecting selection of culture, recreation and sports, factors affecting participation in leisure activities, attitudes towards the supply of cultural services in various localities and the effects of a shorter work week on leisure behaviour patterns. The sample for this stage included 1189 Jewish respondents in 35 localities and 253 Arab respondents in 25 localities. Data on the subjects were collected through interviews. Data on the supply of cultural services were gathered from advertisements in local newspapers.

In the second stage a comparative survey was conducted, in order to assess the changes that had occurred since 1970 in patterns of leisure consumption, the content of leisure activities and values and attitudes related to cultural consumption. Data were gathered from a random, representative sample of 2956 Jewish respondents aged 20 and over. Data on the supply of cultural services were also collected through interviews.

In the third stage a follow-up study was conducted to detect seasonal changes in cultural consumption, recreational habits and tourism behaviour in Israel and abroad. The representative sample included 1197 subjects aged

20 and over in 35 localities and the data were collected by means of interviews.

Partial results of the 1990 survey, together with findings from the 1970 survey for purposes of comparison, are shown in Tables 8.1, 8.2 and 8.3.

MAJOR RESULTS OF THE 1990 STUDY

The major results from the 1990 study can be divided into three main parts: time-budget changes, changes in participation and changes in values.

Time-Budget Changes, 1970–1990

Respondents were requested to reconstruct all of their activities in time units of 15 minutes over the 24 hours before interview. The data in Tables 8.1 and 8.2 indicate that Israelis spend less time sleeping and resting than they did 20 years ago. There was a decrease in the average amount of time devoted to paid work by those in employment, but a small increase overall, probably reflecting the growth of part-time work and the fact that many more women had entered the paid work force. There was a sharp decline in the amount of time devoted to housework, but there was no decrease in the amount of time spent caring for children. In total, leisure time increased by 1 hour per day as compared with 1970, and by 2 hours on Fridays, the first day of the Israeli weekend.

One-third of all leisure time, some 2 hours per day, is spent watching television (less on weekdays, more on Fridays and Saturdays). There was a slight decrease in the amount of time spent reading newspapers, and a slight increase in social activities. There was also a slight increase in the amount of time spent outside the home on Fridays and Saturdays, but in total, the frequency of going outside the house, especially in the older age group, decreased slightly.

Changes in Participation in Cultural Activities, 1970–1990

Table 8.3 shows that participation in arts activities, such as going to the movies, theatre, museums, classical music concerts, was either static or declining – both in the number of participants (movies, theatre) and the frequency of participation (movies, theatre, museums, classical music). At the same time, there was an increase in participation in many 'light' activities, such as social gatherings, going out to a pub or a restaurant, trips in Israel and abroad, active sports, record playing and video watching. Nevertheless there also appeared to be a decline in participation in light entertainment, including listening to popular singers and sports spectating. Possible avenues of interpretation for these changes are associated with: (i) a general increase in hedonistic values; (ii) the search for less collective and more intimate ways of spending leisure time; (iii) the popularity of television and the slight increase in the tendency to remain at home; (iv) a search to replace television viewing with more interactive activities; and (v) the increase in the legitimacy of alternative ways of spending leisure time (Table 8.3).

Table 8.1. Time (in hours) devoted to primary activities on an average weekday/Friday/Saturday, Israel, 1970–1990.

	Weekday		Friday		Saturday	
	1970	1990	1970	1990	1970	1990
1. Gainful employment	4.3	4.5	3.6	2.4	0.5	0.9
2. Household care	2.3	1.3	2.7	2.1	1.1	0.9
3. Shopping	0.4	0.5	0.3	0.7	0.0	0.1
4. Child care	0.5	0.7	0.6	0.7	0.6	0.5
5. Sleep	7.6	7.4	7.3	7.3	9.3	8.9
6. Rest	1.2	1.0	1.4	1.6	1.9	1.4
7. Eating	1.3	1.0	1.6	1.4	1.7	1.4
8. Personal care	0.8	0.8	1.0	0.9	0.8	0.7
9. Prayer	0.1	0.1	0.3	0.2	0.7	0.5
10. Studies	0.3	0.4	0.2	0.1	0.2	0.2
11. Clubs, organizations	0.0	0.1	0.0	0.1	0.1	0.1
12. Newspapers & periodicals	0.4	0.3	0.6	0.4	0.5	0.4
13. Books	0.3	0.2	0.2	0.2	0.2	0.3
14. Radio	0.2	0.1	0.1	0.1	0.2	0.1
15. Television	0.9	1.7	0.9	2.3	0.8	2.0
16. Social life*	1.0	1.2	1.2	1.4	2.1	2.5
17. Conversation	0.2	0.2	0.2	0.2	0.4	0.2
18. Walks	0.2	0.3	0.2	0.4	0.5	0.5
19. Sports	0.0	0.1	0.1	0.1	0.1	0.2
20. Hobby, creative activity	0.1	0.2	0.0	0.1	0.1	0.1
21. Excursions, pleasure trips	0.6	0.4	0.4	0.6	1.3	1.1
22. Other non-home recreation†	0.2	0.2	0.2	0.3	0.3	0.4
23. Work trips	0.6	0.5	0.5	0.3	0.1	0.2
24. Other	0.5	0.8	0.4	0.1	0.5	0.4
Total	24.0	24.0	24.0	24.0	24.0	24.0
Total leisure time	4.1	5.0	4.1	6.2	6.6	7.9
Unweighted *n*	1614	1021	320	168	267	257

* Includes visiting, hosting, parties, dances, games.
† Includes movies, theatre, museums, exhibitions, concerts, light entertainment, coffee houses, night clubs, discotheques.
Source: Katz *et al.*, 1992, weighted data, adults 18 years and over.

Changes in Values

Along with the changes mentioned above, changes occurred in the values
held by Israeli society, or more precisely, there were changes in the emphasis
placed on certain areas, with direct implications for leisure. Some of the
changes which were observed when comparing the 1970 data with the 1990
data are discussed below.

Table 8.2. Time (in hours) devoted to primary activities: average weekday, by gender and
employment status, Israel, 1970–1990.

	Men employed		Women employed		Full-time home–child care	
Activity	1970	1990	1970	1990	1970	1990
1. Gainful employment	7.4	6.0	5.7	4.3	0.2	0.8
2. Household care	0.6	0.5	2.0	1.7	4.9	2.9
3. Shopping	0.1	0.3	0.3	0.5	0.8	0.8
4. Child care	0.3	0.4	0.6	0.9	1.0	1.4
5. Sleep	7.2	7.1	7.4	7.4	8.0	8.1
6. Rest	0.9	0.9	0.9	1.0	1.5	1.4
7. Eating	1.4	1.0	1.1	1.0	1.5	1.4
8. Personal care	0.9	0.8	0.9	0.8	0.8	0.7
9. Prayer	0.2	0.3	0.0	0.0	0.0	0.1
10. Studies	0.2	0.4	0.2	0.2	0.0	0.1
11. Clubs, organizations	0.1	0.1	0.0	0.0	0.0	0.0
12. Newspapers & periodicals	0.4	0.3	0.3	0.2	0.3	0.2
13. Books	0.2	0.2	0.4	0.2	0.2	0.2
14. Radio	0.2	0.1	0.1	0.1	0.1	0.0
15. Television	1.0	1.9	0.8	1.5	1.1	2.0
16. Social life*	0.8	1.1	1.0	1.3	1.2	1.5
17. Conversation	0.2	0.2	0.1	0.2	0.2	0.2
18. Walks	0.1	0.2	0.1	0.2	0.2	0.6
19. Sports	0.0	0.1	0.1	0.1	0.0	0.0
20. Hobby, creative activity	0.0	0.1	0.1	0.1	0.2	0.1
21. Excursions, pleasure trips	0.4	0.6	0.5	0.5	0.8	0.3
22. Other non-home recreation†	0.2	0.2	0.3	0.2	0.1	0.1
23. Work trips	0.9	0.7	0.9	0.5	0.1	0.1
24. Other	0.3	0.5	0.2	1.1	0.8	1.0
Total	24.0	24.0	24.0	24.0	24.0	24.0
Total leisure time	4.0	5.1	4.0	4.6	4.4	5.2
Unweighted *n*	647	248	215	252	359	119

* Includes visiting, hosting, parties, dances, games.
† Includes movies, theatre, museums, exhibitions, concerts, light entertainment, coffee houses,
night clubs, discotheques.
Source: Katz *et al.*, 1992, adults aged 18 and over.

1. There appears to be an increase in the importance of 'present orientation' as compared with 'future orientation'. In 1990, a higher percentage agreed with the statement that 'life is short and dangerous and one should think primarily about the present'.

2. There was an increase in the importance attributed to leisure time, as compared with the importance attributed to work. In 1990, the majority of respondents (67%) attributed equal importance to leisure and work, compared with 48% in 1970. The feeling of a lack of time is felt primarily among the groups with higher education as well as among the young age groups.

3. There was an increase in the value placed on the individual. An examination of the importance attributed to various needs indicated an increase in the importance attached to personal needs, both cognitive (such as 'receiving useful information on day-to-day matters', or 'the desire to study and enrich

Table 8.3. Participation in leisure activities, Israel, 1970 and 1990.

Activity	% Participation (in 24 hours)			Rank		
	1970	1990	Difference	1970	1990	Difference
1. Radio	98	96	−2	1	1	0
2. Television	91	94	+3	2	2	0
3. Social meetings	89	94	+5	3	3	0
4. Newspapers	86	90	+4	4	4	0
5. Excursions, trips in Israel	80	88	+8	5	5	0
6. Cinema	79	61	−18	6	11	−5
7. Book reading	78	77	−1	7	7	0
8. Prayer	74	64	−10	8	9	−1
9. Gambling, lottery	65	56	−9	9	13	−4
10. Theatre	64	49	−15	10	14	−4
11. Museums, exhibitions	63	63	0	11	10	+1
12. Hobbies	60	74	+14	12	8	+4
13. Lectures	56	40	−16	13	19	−6
14. Periodicals	53	45	−8	14	16	−2
15. Records	50	88	+38	15	6	+9
16. Singers, bands	50	49	−1	16	15	+1
17. Spectator sports	31	19	−12	17	22	−5
18. Clubs, organizations	30	18	−12	18	23	−5
19. Studies	28	33	+5	19	20	−1
20. Sports (active)	26	45	+19	20	17	+3
21. Pubs, night clubs	23	42	+19	21	18	+3
22. Trips, tourism abroad	23	57	+34	22	12	+10
23. Reading imported newspaper, journal	21	16	−5	23	24	−1
24. Concerts	19	19	0	24	21	+3

Source: Katz *et al.*, 1992.

oneself') and affective (such as 'being entertained', or 'going out with friends'). On the other hand, there was a moderate decline in the importance placed on collective needs, such as 'to have faith in our leaders', or 'to feel that I am participating in actual events'. There appeared to be an increase in the importance of self-cultivation, along with a decrease in a collectivistic orientation, which finds expression in decreased participation in public and civic activities. It should be noted, however, that the importance attributed to the collective is still high, and the need 'to feel proud that we have a State' still receives a high rating, even though the proportion attributing importance to it decreased slightly.

4. As with the civic/collective orientation, there was a certain decrease in the importance attributed to religious tradition. A comparison of 1970 and 1990 attitudes of the public to the character of the Sabbath indicates an increasingly secular approach. Nevertheless, Israeli society continues to cling to a traditional lifestyle; the importance of family relationships is rated in first place, corresponding to the importance of taking pride in the State. A high percentage observe traditional ceremonies and customs, such as lighting the Sabbath candles or holding a festive family meal on the Sabbath eve, even though a considerable proportion of these view themselves as secular and do not define these customs as religious.

Summary

How does the Israel of 1990 differ from that of 1970? A time-budget analysis is one way of answering this question. There is no doubt that this analysis does not offer a completely satisfactory answer. However, it can provide an indication of a number of significant changes in the day-to-day lifestyle.

The most outstanding change that was expected was that, as a result of the shortening of the standard work week, there would be a significant decrease in the amount of time devoted to work and a corresponding increase in leisure time. This increase would obviously create expectations, hopes and conjecture regarding ways to take advantage of this free time. However, for the population as a whole, the general data indicate that there was no decrease in the amount of time devoted to paid employment except for Friday, the extra day of leisure. These data, however, are somewhat misleading, as the apparent stability in work time is due to more women entering the work force, which involves a decrease in the percentage of the population that is not in paid employment.

What form did the added leisure time take? The only answer available is the increase in the amount of time spent watching television. However, since this increase is also partly due to an increase in the number of viewers, it is difficult to regard this finding as the answer to the question of how additional leisure time was spent. Nevertheless, since the reduction of work hours means a reduction of 5–6 hours in the work week, the daily increase of one hour in the amount of time spent watching television provides a quantitative answer to the above question. If additional time became available as a result of the reduction in the number of work hours, it was 'dissolved' in a variety of different activities, without being channelled in a directed, significant manner to cultural, educational or even family activities.

This conclusion is somewhat disappointing for those who expected, or hoped for, an increased channelling of time in cultural or creative directions. It appears that stability in leisure activities is stronger than attempts to change it. A population that is used to spending leisure time in a particular way is apparently incapable of creating new activity patterns towards which to direct new time resources. The principle of 'more of the same' appears to govern time devoted to leisure.

It should nevertheless be remembered that the methodology used in a time-budget survey is not sufficiently sensitive to changes in time budgeting of individuals or of relatively small groups, and especially not to changes in activities which are not performed on a daily basis (such as going to the movies or to stage performances). Small changes of this type, such as spending time on studies or sports, a hobby, or even spending time in a shopping centre or on more extended social activity, can have great social and cultural significance.

Government Policy Issues

The driving force behind the 1970 leisure study was the Ministry of Education. A dialogue between researchers and policy makers followed the presentation of its results. Two parliamentary committees (Labour, and Education and Culture) devoted several sessions to close inspection of the results, which led to policy making in the area of a shorter work week and a policy on culture and the arts. The results were also examined by the national *Council for Culture and the Arts* and by several national associations in particular areas, such as adult education and sports. The specific issues raised by the different policy makers included the following: whether there is room or need for an explicit cultural policy in a democratic society; how a small nation such as Israel should address the issue of shaping a national identity and blending disparate ethnic cultures into a national culture; how Israel should cope with modernization and the blending of old and new, religious tradition and secular moderniza- tion; whether the mass media can avoid coming under the overwhelming influence of the international free flow of information which inevitably gives an advantage to the politically and culturally strong nations; whether Israel should become a society of European culture, arts and values or a blend of many cultures; what the population should do with the increase in leisure time; what cultural opportunities should be offered with the increased free time and what strategies should be used for leisure education within the school and community systems; how cultural policy makers might help the present less well-educated generation to overcome constraints of age; and how equal opportunities might be guaranteed in the provision and consumption of leisure, in educating towards it, and in the democratization of the arts and access to them.

An effort to formulate leisure policies for Israel was made in 1979, when a group of theorists, practitioners and policy makers participated in an international seminar organized by the Israel Leisure and Recreation Association. The seminar's proceedings (Ruskin, 1984) proposed a platform of recommendations for leisure policies, based primarily on the data presented by Katz and described above. Since then, many of the recommendations have been implemented through a long series of workshops with government practitioners, through numerous publications and through use of the media.

YOUNG PEOPLE AND LEISURE

In addition to the national leisure surveys, which explore the leisure patterns of the whole population, certain attention has been given to investigating the leisure behaviour and preferences of different age groups. Such investigations shed more light on the ways particular age groups spend their leisure and their attitudes towards leisure activities. Three studies on the leisure of youth and children in Israel are described below.

Leisure and Youth

In 1984 a study was carried out to explore the way youth in Israel spend their leisure time, their attitudes towards leisure activities and their preferences (Sivan, 1984). In order to obtain a comprehensive picture of the leisure activities, attitudes and preferences of youth, the research was based on samples of students representing the four main educational frameworks in Israel. These are religious and non-religious schools in the kibbutz (collective village) and religious and non-religious schools in the city. Data were collected using a questionnaire which was submitted to 384 students of junior class in eight schools selected randomly from a list of secondary schools supplied by the Ministry of Education. The questionnaire included a list of activities which were divided into groups, based on theoretical background and on preliminary interviews conducted with young people to elicit information on additional leisure activities engaged in by youth in Israel. The groups comprised instrumental, expressive, religious (performed alone and in group), and voluntary activities.

Results showed a strong tendency to prefer and to participate in expressive leisure activities such as watching television, listening to pop music, meeting with friends, participating in parties and travelling. The average frequency of participation in these activities was between one and three times a week. The frequency of such activities was much higher than the frequency of instrumental activities such as taking study courses or doing voluntary work, which were performed by young people only once or twice a month. Examination of the results revealed a significant difference between youth from different educational backgrounds. There was a strong tendency among youngsters from the kibbutz to participate in social leisure activities and for youngsters from religious schools to participate more in religious activities (Sivan, 1986). Both the kibbutz and the religious environments seemed to influence the level of participation in voluntary activities. The community life of the kibbutz, which is a unique way of life in Israel, was found to be an influential factor in the way youth spend their leisure time. Follow-up research is planned to explore any possible changes in the way Israeli youth spend their leisure time and to examine changes in their leisure patterns.

Extracurricular Activities of High School Students

Another survey was carried out by the Central Bureau of Statistics during the 1990/91 school year. The survey was on extra-curricular activities of ninth to twelfth grade students in the Hebrew and the Arab education sectors (Ministry of Education and Culture, 1993). It was based on a sample of classes which were studied as a whole for the first time. The sample of 12,300 respondents was drawn up in two stages: in the first stage a representative sample of schools was selected according to various characteristics of the institution and second, a representative sample of classes was selected in which each of the grades participating was represented. The survey was carried out within classes where each student was required to fill out a questionnaire.

The questionnaire included questions on activities such as at home, away from home and after school activities, activity in youth movements, activity in community centres, and voluntary activities. In addition the questionnaire included demographic data and a series of hypothetical questions on preferred Friday morning (weekend) activities, assuming no school on Friday.

Results showed that, in both Hebrew and Arab sectors, watching TV was the most popular home activity. Watching video films was also popular among both sectors. About one-third used personal computers at least once a week and the main use was for playing games. The majority of both sectors read at least one newspaper a week and at least one book a month.

Going to the cinema was the most popular 'away from home' activity among the Hebrew sector and attending sports competitions was the most popular among the Arab sector. Sports activities were the most popular after school activities in both sectors. More than one-third of the students in both sectors participated in youth movements in the course of a year. About one-third participated in voluntary activities and more than one-third participated in summer camps.

In both sectors there were differences in participation rate for activities between males and females. Males watched TV and videos and played with computers to a higher extent than females, whereas females were involved more in reading books and magazines. Overall in the Arab sector males participated more in 'away from home' social activities whereas in the Hebrew sector there were differences in participation rate in different activities. There were more differences in participation rates in activities between different age groups in the Hebrew sector than in the Arab sector.

Leisure and Children

A further source of additional information on leisure of children and youth in Israel is the Research and Information Centre of the National Council for Children's Welfare, which publishes a statistical annual on the subject of *Children in Israel* (Ben-Arie, 1992).

The report contains a chapter on 'Children and the World of Leisure', which gathers information from studies of leisure behaviour patterns conducted by the Central Bureau of Statistics, studies of the Advertisers' Association, which gathers information on exposure to the media, and also data from the Community Centres Corporation of Israel and the Research and Information Centre of the National Council for Culture and the Arts. These articles are not comprehensive, but each contains separate data regarding leisure of children and youth to age 18, such as reading books, sports and social activities, activities in community centres, cultural and arts activities, movie going and video viewing and the number of visits to the movies.

From these reports, there emerges a picture of considerable leisure participation among Israeli children and youth, primarily as a result of the wealth of social activities and informal educational activities offered to members of these age groups. A comprehensive survey of leisure of school children in Israel was planned by the Central Bureau of Statistics for the year 1993.

REFERENCES

Ben-Arie, A. (1992) *Children in Israel – Statistical Annual, 1992*, Research and Information Centre, National Council for Children's Welfare, Tel Aviv.

Katz, E. and Gurevitch, M. (1976) *The Culture of Leisure in Israel: Patterns of Spending Time and Consuming Culture*, Tel-Aviv, Am-Oved: Culture and Communication in Israel, Faber and Faber, London.

Katz, E. (1994) Problems of leisure and culture in a new nation: the Israeli experience. In: Ruskin, H. (ed.) *Leisure: Towards a Theory and Policy*, Proceedings of the International Policies, Jerusalem 1979. Fairleigh Dickenson University Press, Teaneck, NJ.

Katz, E. *et al.* (1992) *The Culture of Leisure in Israel: Changes in Patterns of Cultural Activity, 1970–1990*. Israel Institute of Applied Social Research, Jerusalem (Hebrew).

Ministry of Education and Culture, Division for Social and Youth Education (1993) *Extracurricular Activities of 9th–12th grade pupils, Hebrew and Arab Education 1990/91, Part I, Special Series No.946*. Central Bureau of Statistics, Jerusalem.

Ruskin, H. (ed.) (1984) *Leisure: Toward a Theory and Policy*, Proceedings of the International Seminar on Leisure Policies, Jerusalem 1979. Fairleigh Dickenson University Press, Teaneck, NJ.

Sivan, A. (1984) *Leisure of High-School Youth in the Israeli Kibbutz and City*. Bar-llan University, Ramat-Gan, (Hebrew).

Sivan, A. (1986) *Influences of Beliefs and Values on Leisure of Youth*. Multi-Purpose Instructional Centre, Haifa.

9 Japan

MUNEHIKO HARADA

YEARS OF EXPONENTIAL GROWTH

In Japan, in the latter half of the 1970s and the first half of the 1980s, high growth in the leisure market was stimulated by general economic development. The Y3.9 billion leisure market of 1982 approximately doubled to Y7.5 billion by 1992. This period has been referred to as the *years of exponential growth* (Harada, 1994). Exponential growth means that the economic growth rate in a given year exceeds the preceding year's growth rate, which in turn accelerates the increase in scale of the market. The period was also characterized by a substantial change in Japanese behaviour, in the form of a shift toward a more leisure-orientated lifestyle.

With the increased awareness of leisure-orientated lifestyles, the leisure market experienced changes not only in size but also in substance. According to the 1983 *Public Census on Lifestyles* (Prime Minister's Office, 1983) *leisure* was considered, for the first time, to be the most important aspect of daily life, while 'eating habits' and 'living habits' decreased in importance compared with previous years' polls. The percentage of survey respondents placing *leisure* in the highest importance category has increased every year since.

With the twenty-first century drawing near, the Japanese government has recently been given the task, in line with public demands, of creating a leisure-orientated environment and providing more opportunities for leisure enjoyment. In the latter half of the 1980s, the issue of shortening working hours was placed on the national policy agenda, in the interest of improving the working conditions of the Japanese worker. Demands for shortened working hours increased. The results came in 1992, when general annual labour hours were officially reduced to 1972, falling symbolically below 2000 hours for the first time. Additionally the Resorts Act was passed, which relaxed environmental regulations and provided structures for subsidy and tax breaks to promote leisure-orientated development in the private sector. This set the stage for Japan's first ever resort boom.

As public interest in recreational activities grew, governmental organizations began to study the issue of leisure. Surveys were mounted, including the 1986 *Work and Leisure in the 1980s* survey, conducted by the Economic Planning Agency (1986a, b), and the 1991 *Public Opinion Poll on Leisure and Travel* and the *Public Opinion Poll on Working Hours and the 2-Day Weekend*, conducted by the Prime Minister's Office (1991a, b). These were, however, partial polling at best, since samples were drawn from limited areas and organizations, such as businesses; they were not nationwide and did not involve any follow-up research. In addition, in 1993, a non-profit organization, the Sasagawa Sports Foundation, surveyed some 2000 persons aged 15 and over across the nation, but interviewed only those persons actively participating in sport, so failed to represent the leisure situation of the Japanese population as a whole (Sasagawa Sports Foundation, 1993).

NATIONWIDE STUDY ON LEISURE AND RECREATION ACTIVITIES

The *Study on Recreational Activities* is a regular nationwide study on leisure in Japan, begun in 1977 by the Leisure Development Centre, an extra-governmental body of the Ministry of International Trade and Industry (MITI). From the data drawn from the study, this organization publishes an annual *White Paper*, which reports on the leisure and recreational activities of the Japanese people and trends in leisure related industries (Leisure Development Centre, 1995). The study polls 4000 men and women aged 15 and over from across the nation. Initially the survey involved a sample of 3000 from urban areas only. From 1987, each year some 3000 persons are randomly sampled from urban areas with a population of 50,000 and over, and 1000 persons are randomly sampled from urban and rural areas with populations of less than 50,000. The survey response rate has exceeded 80% each year, suggesting that the survey provides a fair representation of Japanese public opinion.

The survey categorizes 89 leisure activities into four groups: (i) sports (27 items); (ii) hobbies and entertainment (30 items); (iii) games and amusements (20 items); and (iv) travel and tourism (12 items). The published results indicate levels and frequency of participation, expenditure and desired activities. Some of the key data generated are as follows:

- *percentage participation*: percentage of respondents taking part in each leisure activity one or more times in a year;
- *frequency*: average number of times participants take part in each leisure activity in a year;
- *expenditure*: average amount of money per capita spent in taking part in each leisure activity in a year;
- *desired participation*: percentage of respondents wishing to take part in or wishing to continue participating in each leisure activity.

CHARACTERISTICS OF LEISURE ACTIVITIES IN JAPAN

Table 9.1 shows participation rates and frequency of participation for the 89 activities in 1982 and 1992. Although figures in the Table are not classified by gender, health conscious sports, such as jogging, long-distance running (marathons), exercise, bowling and swimming and, to a lesser extent, tennis and skiing, were popular among both men and women, while golf was more popular among men. However, with the exception of skiing, soccer, swimming and bowling, participation in these activities declined or was static between 1982 and 1992. There are two possible explanations for this trend. Firstly, some activities may have been affected by the recent recession. For example, golf (at the golf course) and golfing practice (at the driving range) showed steady growth in participation rates up to 1991, which marked the start of the collapse of the 'bubble' economy. Secondly, more leisure activities were available to the general public in 1992 compared with 1982. Because of the diversification of leisure opportunities in recent years, such outdoor sports as skin/scuba diving, surfing/windsurfing, yachting/motorboating and hang/para-gliding were added to the list of activities in 1992. It could be conjectured that, as people had more opportunities to become socialized into newly introduced outdoor sports, participation rates in traditional sports tended to decline.

Under hobbies and entertainment, watching videos, listening to music, gardening and horticulture, watching movies, watching sports, playing music and attending concerts were popular among Japanese. Compared with 10 years earlier, more interest was shown in video and music in 1992, whereas participation in gardening and horticulture and movie going were down. The appearance of the rental video business and the arrival of the new compact-disc medium are likely reasons for increased video watching and music listening respectively.

Among games and amusements, the popularity of dining out and karaoke, lotteries, drinking, and board and card games figured high among Japanese. Of these, the level of participation in karaoke entertainment was particularly high, showing an increase of some 20% between 1982 and 1992.

Table 9.1. Leisure activity participation and frequency, Japan, 1982, 1992.

Sports	% participating in year			Average annual frequency		
	1992 (n = 2414)	1982 (n = 1606)	Change 1982 to 1992	1992	1982	Change 1982 to 1992
Jogging/marathons	23.6	28.5	-4.9	38.0	39.7	-1.7
Exercise	29.5	42.2	-12.7	55.6	60.4	-4.8
Physical training	12.1	12.1	0.0	45.4	47.4	-2.0
Aerobics/jazz dancing	4.9	4.2	0.7	26.0	30.7	-4.7
Table tennis	13.6	18.9	-5.3	14.1	16.3	-2.2
Badminton	14.7	20.0	-5.3	11.8	15.0	-3.2
Playing catch/baseball	20.2	25.5	-5.3	16.9	22.4	-5.5
Softball	13.9	18.3	-4.4	10.4	12.6	-2.2
Cycling	15.1	14.2	-0.9	26.8	27.0	-0.2
Ice skating	7.9	8.6	-0.7	3.6	3.5	0.1
Bowling	36.0	29.4	6.6	6.5	6.4	0.1
Soccer	6.8	4.2	2.6	20.6	20.2	0.4
Volleyball	11.7	14.4	-2.7	18.7	20.5	-1.8
Basketball	6.9	7.6	-0.7	26.2	22.6	3.6
Swimming	25.5	23.2	2.3	13.6	10.4	3.2
Japanese martial arts	3.3	3.9	-0.6	42.0	34.6	7.4
Gateball croquet	1.4	2.4	-1.0	36.0	32.2	3.8
Golf	14.3	12.3	2.0	11.7	16.8	-5.1
Golfing practice	18.8	–	–	18.5	–	–
Tennis	13.4	15.2	-1.8	19.6	24.1	-4.5
Riding	0.6	0.5	0.1	6.5	22.0	-15.5
Skiing	17.9	11.3	6.6	5.5	5.7	-0.2
Fishing	17.0	18.2	-1.2	9.1	11.1	-2.0

Skin/scuba diving	1.0	—	—	5.8	—	—
Surfing/windsurfing	0.7	—	—	10.4	—	—
Yachting, motorboating	1.4	—	—	4.6	—	—
Hang/para-gliding	0.3	—	—	24.1	—	—
Hobbies and entertainment						
Writing (novels, poems, etc.)	6.5	8.7	-2.2	18.9	20.6	-1.7
Photography	9.5	12.9	-3.4	12.4	11.5	0.9
Making/editing video tapes	6.8	4.1	2.7	11.9	—	—
Watching video tapes (including rental tapes)	43.5	—	—	22.0	9.5	12.5
Choral singing	2.8	5.0	-2.2	29.3	18.3	11.0
Playing western music	7.7	9.0	-1.3	51.4	54.0	-2.6
Japanese traditional music	3.2	6.3	-3.1	45.5	44.6	0.9
Drawing/painting/sculpting	7.8	9.1	-1.3	19.3	21.8	-2.5
Ceramics	2.2	2.0	0.2	17.5	8.6	8.9
Handicrafts	4.7	6.3	-1.6	22.8	15.6	7.2
Model-building	4.5	6.8	-2.3	11.5	9.0	2.5
Do-it-yourself	12.0	17.1	-5.1	11.4	12.2	-0.8
Gardening/horticulture	31.6	38.1	-6.5	39.3	44.4	-5.1
Knitting/weaving	17.4	28.7	-11.3	27.1	29.1	-2.0
Sewing	12.8	21.9	-9.1	28.8	32.3	-3.5
Cooking (except regular meals)	11.8	15.0	-3.2	15.0	20.4	-5.4
Going to athletic events	22.0	20.0	2.0	5.6	6.6	-1.0
Going to movies	31.1	34.5	-3.4	5.2	5.5	-0.3
Going to theatre	14.2	14.2	-0.1	3.7	4.0	-0.3
Going to artistic performances	5.2	6.8	-1.6	3.2	4.0	-0.8
Going to concerts	22.9	19.9	3.0	4.3	5.0	-0.7
Listening to music	42.0	35.0	7.0	69.6	72.1	-2.5
Going to art exhibitions	16.8	13.7	3.1	5.8	7.0	-1.2

Table 9.1. *continued*

	% participating in year			Average annual frequency		
	1992 (n = 2414)	1982 (n = 1606)	Change 1982 to 1992	1992	1982	Change 1982 to 1992
Hobbies and entertainment						
Calligraphy	7.3	7.1	0.2	33.6	36.9	−3.3
Tea ceremony	4.3	5.5	−1.2	22.5	28.2	−5.7
Flower arranging	6.8	8.1	−1.3	30.3	31.5	−1.2
Traditional Japanese dance	1.4	2.2	−0.8	38.0	42.4	−4.4
Western dance (e.g. ballroom)	1.9	3.6	−1.7	29.9	21.7	8.2
Personal computer networking	1.7	–	–	46.0	–	–
Non-work study and research	14.2	12.9	1.3	39.9	39.4	0.5
Games and amusements						
Go	4.9	12.3	−7.4	28.4	22.5	5.9
Shogi	12.0	24.7	−12.7	15.6	19.4	−3.8
Board and card games	36.5	50.9	−14.4	12.5	13.5	−1.0
Karaoke	51.5	30.1	21.4	11.4	14.8	−3.4
Video games	27.1	27.1	0.0	38.6	20.4	18.2
Video arcade games	22.7	22.0	0.7	13.8	14.9	−1.1
Mahjong	13.3	23.3	−10.0	13.2	20.8	−7.6
Billiards	6.8	4.9	1.9	6.0	6.8	−0.8
Pachinko	27.7	34.0	−6.3	25.7	20.5	5.2
Lotteries	38.0	–	–	5.4	–	–
Horse races (JRA*)	11.5	8.8	2.7	13.4	13.6	−0.2
Horse races (NAR†)	3.2	–	–	12.1	–	–
Bicycle races (spectator)	1.5	2.1	−0.6	7.5	11.1	−3.6
Boat races (spectator)	1.7	2.4	−0.7	21.7	11.8	9.9

Auto races (spectator)	0.4	0.8	−0.4	20.8	7.8	13.0
Eating out (except regular meals)	68.8	61.1	7.7	17.4	18.8	−1.4
Visit bars/pubs	44.1	42.4	1.7	16.6	18.0	−1.4
Clubs and cabarets	5.6	10.6	−5.0	7.7	11.2	−3.5
Discotheques	3.6	7.5	−3.9	8.9	8.1	0.8
Public saunas	10.9	8.8	2.1	13.6	10.4	3.2
Travel and tourism						
Amusement parks	39.9	37.6	2.3	3.6	4.1	−0.5
Going for drives	56.3	52.3	4.0	12.5	10.6	1.9
Picnics, hikes, nature walks	36.4	39.2	−2.8	9.1	–	–
Mountain climbing	8.5	–	–	3.9	–	–
Camping	5.1	–	–	2.3	8.7	−6.4
Field athletics	6.5	10.5	−4.0	3.3	2.7	0.5
Going to the beach	31.0	38.6	−7.6	2.8	2.7	0.1
Zoos, botanical gardens, aquariums, museums	44.0	38.4	5.6	3.3	3.3	0.0
Exhibitions and events	26.3	26.1	−0.2	3.9	4.0	−0.1
Family reunions	23.6	24.4	−0.8	3.5	3.6	−0.1
Domestic tourism	60.6	54.4	6.2	3.1	3.0	0.1
Overseas trips	10.8	6.1	4.7	1.4	1.4	0.0

Source: Prime Minister's Office, 1991a.

* JRA = Japan Racing Association.

† NAR = National Association of Racing.

Note: to compare the 1992 data with the 1982 data, for 1992 only urban areas sample of 3000 are included.

This boom has not affected Japan alone, having spread throughout the Pacific rim to China, Taiwan, Korea, Australia and New Zealand.

Finally, in the travel category, participation levels in domestic sightseeing and vacationing, going for drives, and visiting zoos, botanical gardens, public aquariums, museums and amusement parks were all high, and increased between 1982 and 1992. In contrast, traditional forms of vacationing activity, such as going to the beach or family reunions, declined in popularity. The reduction in seaside trips can be attributed partly to loss of natural coastline, but also to the fact that many newly built, large indoor swimming pools may be luring people away from the beach.

GOVERNMENT POLICY AIMED AT THE PRIVATE SECTOR AND INCREASED RECREATION PARTICIPATION

The period from the latter half of the 1960s into the 1970s was associated with a growth in facilities which provided opportunities for people to take part in leisure activities. In particular, large-scale development of golf courses and resort hotels became profitable enterprises because of governmental privatization of land and a general willingness among businesses to invest capital in the so-called 'bubble economy'. However, the results of this government and business activity were not reflected in increased participation levels, as shown by the official surveys. In later surveys, activity categories were increased to include watching videos, computer-based activities, hang-gliding and para-gliding, and camping, while surfing, sailing and skin-diving were listed separately rather than as one activity category as before. This increase in activity categories, aimed at tracking leisure trends more precisely and comprehensively, is believed to be the reflection of the diversification of recreational opportunities.

SHORTENED WORKING HOURS AND GREATER DIVERSIFICATION IN RECREATIONAL BEHAVIOUR

In the 1960s and 1970s Japanese workers were working an average of 2300 hours per year and the 2-day weekend was practically non-existent in the business sector. Thus few had time for much leisure, and recreational behaviour was limited. It has been noted that the Japanese government took steps to reduce normal working hours from 2111 hours a year in 1988 to 1972 hours in 1992. The expressed need for, and effects of, shortened working hours and greater free time can be read from the results of the surveys. In the 1980s, personal preferences appeared to shift away from materialistic concerns to embrace more humanistic values and a desire for a better quality of life. Data on desired leisure participation reveal that, in 1992, the most desired activities were overseas travel (36.8%), domestic sightseeing and vacationing (16.4%) and camping (13.2%). Moreover, a change can be seen in leisure in general, as greater proportions of people were taking part in a more extensive array of activities.

LEISURE AND SOCIAL INEQUALITY

The Japanese are, basically, extremely homogeneous in their characteristics. The surveys examined in this chapter unfortunately were not designed to ascertain respondents' socioeconomic characteristics, such as education, religion, income or class. However, according to a 1991 survey conducted by the Prime Minister's Office (1991a), 90% of Japanese working people considered themselves to be part of the 'middle class'. For this reason, the issue of self-reported social class and leisure participation is difficult to assess in Japan. In the Edo (1603–1867) and Meiji (1868–1911) eras, leisure was reserved for the privileged classes of *samurai* and the nobility, but the leisure boom phenomenon of modern day Japan seems to involve the entire nation being drawn to the golf course or karaoke lounge. Social inequality with regard to leisure seems to be related more to availability or lack of free time than to level of income. However, it is clear that the 'leisure repertoire' has expanded significantly for the mass of the people compared with the past.

Despite the official reduction in working hours, the free time of the average working male, aged 30 to 50, living in an urban area, is still greatly limited by long working hours and the great amount of time spent commuting to and from work. Some 60% of such individuals spend 1 hour or more each way, in commuting to work, often returning home after 8 p.m. in the evening. Although the majority of Japanese view themselves as belonging to the 'middle class', in reality some, because of their low incomes, belong to the 'lower class'. Lower class workers inevitably have to work longer hours to make up for lower income levels, while others, either because of their higher occupational position or simply out of habit, sacrifice their free time to put in more working hours. These circumstances create a different situation from that described by Schor (1991) who, in describing the *Overworked American*, wrote that '... inequality of income creates inequality of time'. She speaks of American workers having sufficient money to enjoy themselves, but not having the leisure time in which to spend it, making it totally impossible to describe them as a 'leisure class'.

A PROBLEM OF METHODOLOGY

The problems of nationwide surveys on participation, as Kelly (1980) pointed out, are that: (i) studies aimed at formulating theory are lacking; (ii) variations in levels of participation are partly related to the supply of facilities; and (iii) the uncertainty of reported 'desired' activities demands the formulation of theory relating to 'substitution' and 'socialization', rather than just measuring participation. These are issues that must be confronted in regard to the future of such surveys.

Participation rates reported here relate to the percentage of persons taking part in a given recreational activity one or more times during the past year, but do not identify 'regular' participants who take part in the activity frequently. The frequency of participation data is presented in aggregate terms only and is not used to distinguish between regular and infrequent participants. It has been shown (Chase and Harada, 1984) that, when using a 1-year reference period, survey respondents tend to overestimate the number of times they take part in an activity. As such, the results of surveys using such a reference period may be useful in determining trends in participation in leisure activities, but should be used with caution when formulating theory and policy. Furthermore, as Veal (1984) has pointed out, it would be best to improve the validity of the data by conducting such surveys four times a year to obtain a view of seasonal variations in participation.

CONCLUSIONS

The survey on leisure participation which the extra-governmental organization of the Ministry of International Trade and Industry has conducted regularly over the past 16 years, has provided a database on Japanese leisure activity at a crucial time. This database has been an important source of information for making policy decisions, particularly industry-directed policy. The survey has also provided evidence of the leisure and recreation needs of the general population when government decisions were being made on resort development and shorter working hours.

As we approach the twenty-first century, participation by Japanese people in leisure activities is predicted to rise, since the general opinion in the nation continues to support reform of basic social and economic structures. Outdoor recreation activities, particularly highly desired activities such as short break vacations and longer holidays, are expected to grow in popularity. Also, as the 'leisure-able' generation ages, the need to accommodate the demands of the elderly for recreation will become more pressing (Fujimoto and Harada, 1994). The advantage of the surveys reported here is that they take place every year. Taking advantage of this feature of the surveys, personal leisure and socialization processes can be analysed, enabling accurate conclusions to be drawn concerning the ageing of society and changing recreational demand. If applied in this sense, the surveys would, in fact, be providing data that would be useful in formulating theory in relation to such topics as *substitution* and *resocialization*, rather than just measuring levels of participation. This is one issue to be discussed in regard to the future of leisure research.

REFERENCES

Chase, D. and Harada, M. (1984) Response error in self-reported recreation participation. *Journal of Leisure Research* 16 (4) 322–329.

Economic Planning Agency (1986a) *Educating the Importance of Leisure.* Government Printing Office, Tokyo.

Economic Planning Agency (1986b) *Work and Leisure in the 1980s* Government Printing Office, Tokyo.

Fujimoto, J. and Harada, M. (1994) A life course perspective and sports participation patterns among middle-aged Japanese. In: Henry, I. (ed.) *Leisure: Modernity, Postmodernity and Lifestyles: Leisure in Different Worlds Vol. 1.* Leisure Studies Association, Eastbourne, Sussex, pp. 209–216.

Harada, M. (1994) Towards a renaissance of leisure in Japan. *Leisure Studies* 13 (4), pp. 277–287.

Kelly, J.R. (1980) Leisure and quality: beyond the quantitative barrier. In: Goodale, T.L. and Witt, P.A. (eds) *Recreation and Leisure: Issues in an Era of Change.* Venture, State College, PA, pp. 300–314.

Leisure Development Centre (1995) *Leisure White Paper '95* (in Japanese). LDC, Tokyo.

Prime Minister's Office (1983) *The Public Census on Lifestyle.* Government Printing Office, Tokyo.

Prime Minister's Office (1991a) *Public Opinion Poll on Leisure and Travel.* Government Printing Office, Tokyo.

Prime Minister's Office (1991b) *Public Opinion Poll on Working Hours and the 2-day Weekend.* Government Printing Office, Tokyo.

Sasagawa Sports Foundation (1993) *Sports Life Data, Vol. 3.* SSF, Tokyo.

Schor, J.B. (1991) *The Overworked American.* Basic Books, New York.

Veal, A.J. (1984) Leisure in England and Wales. *Leisure Studies* 3 (2), 221–229.

10 New Zealand

ALLAN LAIDLER AND GRANT CUSHMAN

EARLY IMAGES

Systematic research into leisure patterns in New Zealand has a short history. Before the last few decades of the twentieth century, the patterns were obscure and sketchy. The occasional account of economic, social and cultural life before European immigration, as in Best's study of *Games and Pastimes of the Maori* (1976), throws a little light on the obscurity, but most of the culturally important oral histories of Maori life belong to a tradition not widely understood by the nineteenth-century immigrants and, until comparatively recently, not recorded by those who wrote in English.

Social histories of European society in New Zealand from about 1840 reflect a passing interest in leisure, particularly that of the British settlers and their descendants (see Perkins and Gidlow, 1991), but they depict little of the leisure behaviour of other groups: the Chinese and continental Europeans who came early; the large numbers from Asia and the island nations of the south-west Pacific who came later, especially from Western Samoa, Tonga, Fiji and the Cook Islands.

FROM LOCAL TO NATIONAL SURVEYS

A review of early surveys (Jorgensen, 1974) concluded that most had been superficially descriptive and methodologically weak, even though the resulting data and experience proved beneficial in subsequent research. The initial surveys were conducted at local or regional level and reflected the research needs of public administrators and officials aware of expanding urban populations, the increasing use of public land for outdoor recreation, growing affluence and changes in discretionary time and income. Particular attention fell on the leisure-time activities of youth and – more widely – on the implications of sedentary lifestyles.

Public policy makers with an eye on community health and social reform were keen to identify and foster leisure activities that were 'constructive, socially responsible and wholesome' (Cushman, 1993) and encouraged surveys that focused on sport and physical and outdoor recreation and left less vigorous and more informal leisure behaviour, such as home-based activity, in the background. Not surprisingly, and perhaps not only coincidentally, contemporary recreation planning placed a heavy emphasis on sport and physical pursuits outdoors.

Research projects on a larger scale were carried out from the mid-1970s and details of these are given in Table 10.1. While it is not possible here to deal comprehensively with each project, some comparisons can be made. The data were gathered for varying purposes, using differing methodologies, time periods, conceptualizations of leisure and measures of independent and dependent variables, but they contained enough similarities to open up a macroscopic view of leisure patterns in New Zealand.

The *New Zealand Recreation Survey* (NZRS) was carried out in the mid-1970s (Tait, 1984). Its aim was to give more of a national picture than had so far been sketched by extrapolating from small samples. In general terms, New Zealand seemed to favour leisure activities that were mainly sedentary and/or home-centred. Active involvement in sport appeared to be low when activities were looked at one by one, although, when clustered into a single class of activity, the groupings were not insignificant (Robb and Howorth, 1977). Although television watching, sexual activity and pub-going were conspicuously missing from the NZRS, it would be unwise to suppose that these activities had low participation rates. Local knowledge and experience and more recent research suggest otherwise. The items were simply omitted from the survey lists – the first two by design, the third by oversight.

A striking finding of the study was the domestic concentration of leisure activity, with at least eight of the top ten activities being undertaken at home or at a friend's or relative's home. In addition, the NZRS data added to a growing realization that most of the leisure activities typically listed for selection in surveys are undertaken by minorities: 191 out of the listed 216 leisure activities attracted participation from less than 14% of respondents.

While questions on leisure have not yet been included in a New Zealand census, probably the nearest equivalent was a *Social Indicator Survey* (SIS) conducted in 1980–81 by the Department of Statistics (1984). Part of it concentrated on leisure and built a profile similar to that outlined by the NZRS, confirming the attraction of home-based and less physically demanding interests. There is some support in the SIS for the stereotype of New Zealand women as being more interested in, or confined to, activities and social interaction in the home than men, although for the majority of both, visiting and being visited played a notable part in weekly life. There was also some support for the common assumption in New Zealand that the pub and the outdoor sporting scene are predominantly male preserves, although this is certainly less so now than in the 1960s and 1970s.

PROFILE OF A NATIONAL PARTICIPATION SURVEY

Efforts to draw a fuller description of New Zealand at leisure were incorporated into the 1991 *Life in New Zealand* (LINZ) (Wilson *et al.*, 1991) survey. In this, a self-administered questionnaire was sent to approximately 10,000 adult New Zealanders and provided the starting point for a nationwide investigation into a number of health-related factors in lifestyle and behaviour, including physical activity, nutrition, general health and leisure.

Table 10.1. Features of three New Zealand national surveys.

	NZ Recreation Survey (NZRS)	Social Indicator Survey (SIS)	Life in New Zealand Survey (LINZ)
Dates	1975	Oct 1980–Sept 1981	April 1989–May 1990
Population estimates	All New Zealanders 10 years and over	All New Zealanders 15 years and over	All New Zealanders 15 years and over
Sample size	4011	6891	4373
Sampling techniques	Stratified clustered random sample from six 'community types'	Stratified clustered random sample of permanent households	Random samples from electorates and electoral rolls
Survey instrument	Pretested standardized questionnaire. Trained interviewers	Pretested standardized questionnaire. Trained interviewers	Pretested standardized mail-out questionnaire
Major dependent variables	Home and away-from-home activities in preceding 12 months	Home and away-from-home activities in preceding 12 months	Home and away-from-home activities in preceding 4 weeks
Commissioner	NZ Council for Recreation and Sport	Department of Statistics	Hillary Commission for Sport, Fitness & Leisure
Researchers	Market Research NZ and Dept of Internal Affairs	Department of Statistics	Otago, Lincoln and Victoria Universities

Source: Laidler and Cushman (1993).

Sponsors, Researchers and Aims

The survey was shaped with a strong health ethos and ethical approval for the total project was cleared with a regional health authority. The principal funder, the Hillary Commission for Recreation and Sport (later for Sport, Fitness and Leisure), had already declared a concern about the 'outcomes of inactive and otherwise unhealthy lifestyles'. The Ministers of Health and of Recreation and Sport echoed the concern and gave the project their political blessing.

The research was coordinated by staff of the School of Physical Education, University of Otago, who had previously cooperated with the Hillary Commission on a study entitled *The Cost of Doing Nothing* (Russell *et al.*, 1987), following an investigation into New Zealanders' physical activity and inactivity. The LINZ study was much broader and it involved staff from four universities with specialist research interests covering not only health, exercise and nutrition but also leisure and survey methodology.

A number of sponsors with good *prima facie* reasons for being 'health conscious' were found: the Apple and Pear Marketing Board, the National Heart Foundation of New Zealand, the New Zealand Cancer Society, the Medical Research Council, and the Department of Health. Other support came from Philips, Telecom, Avis Car Rentals and Television New Zealand.

The intention of examining patterns of leisure participation was included among a number of aims, most of which reflected the central concern that life in New Zealand could be healthier. In the context of this rationale, active leisure assumed a greater relevance than passive 'inactivity' to the primary aims of the research, which were:

- to gather a database on physical activity patterns against which intervention programmes could be assessed;
- to examine relationships between levels of physical activity and a variety of health related factors of lifestyle;
- to identify physical and social barriers to increased fitness levels.

Data were collected during the period August 1989 to March 1990. Before that, a pilot study had been undertaken to develop instruments, procedures and logistics for the proposed national survey. A report on the pilot was prepared and a paper published on the validity and reliability of some of the data collected (Hopkins *et al.*, 1991). The aim was to include New Zealanders aged 15 years and over. An adult sample was drawn from current electoral rolls but, as 15–18-year-olds were not included in these, a 'snowball' technique was used to generate a sample for this category.

Data were gathered in two phases. In Phase I, the main *Life in New Zealand* questionnaire, which sought demographic details but also touched on eating and activity patterns, was posted each month to ten people randomly chosen from each of the country's 97 electorates. When this was returned, two of the following four questionnaires were mailed to the respondents: *Your Health, Changes in Your Life, Your Eating Habits, Leisure.*

Phase II involved a health check and the completion of questionnaires (as in Phase I) by people selected from 20 electorates representative of the country as a whole, including the North and South islands and rural and urban areas. The health checks included measures of height, weight, skinfolds, blood pressure, heart rate, 24-hour diet recall, cholesterol and triglyceride counts.

Attempts were made to boost the response rate by follow-up telephone calls and by post, by newspaper, radio and national television prompts and by opening data-collection centres in the evenings and on Saturdays. A subsequent analysis of non-respondents (away from address, deceased, refused, etc.) produced a correction factor for the calculation of monthly means. The 'best estimate' of an overall response rate for Phase I was 80% and for Phase II it was 56%.

All data were weighted to compensate for differences in age and sex composition between the study sample and the previous national census of population. Differences in numbers of questionnaires completed monthly were also balanced out in an attempt to remove possible bias in seasonality.

The profile of leisure participation was created with reference to the dimensions of activity, time, place, spending, experience and opinion. Although leisure was examined in a similar research context to patterns of eating, drinking, smoking and exercising, the scope was not narrow. It included home-based leisure, social activities and entertainment, arts and cultural interests, sports and outdoor pursuits, recreational tourism and informal learning. The research was designed to find rates and frequency of participation in favoured activities, reasons for participation and non-participation, reasons for taking up and giving up involvement, facility use, club membership, engagement in voluntary work, leisure expenditure and attitudes to leisure.

RESULTS OF THE LEISURE IN NEW ZEALAND SURVEY

The Big Picture

New Zealanders who answer questions on their leisure in national surveys seem comparable with respondents to similar questions in other countries at a similar stage of industrial and technological advancement, such as Canada and Australia. Activities (and forms of inactivity) variously defined as 'favourite', 'popular' or 'enjoyed' are predominantly sedentary, requiring low levels of physical exertion. Engagement in them usually occurs at or close to home, requiring little formal organization or use of publicly provided facilities.

Specialized skill is often not a prerequisite. The outcome is commonly social interaction, entertainment or the development of personal interests. Although forms and style vary with such factors as family situation, social status and cultural identity, 'the considerable commonality of adult leisure' that Kelly (1990, p.37) has called the leisure 'core', seems to apply in the New Zealand setting as elsewhere, involving a number of relatively accessible, low-cost and often home-based activities that most people do frequently. Some involve the media, especially reading and television; many are social, involving informal interaction with intimates who are most commonly those living in the same household.

Table 10.2 suggests, moreover, that there is stability in the core over time. Although the three surveys referred to varied in research design, instruments and execution, the same activities tended to be most highly ranked in all three. The aggregation of data inevitably obscures fine detail. Table 10.2, for example, does not show the considerable variation in the rankings by gender or socioeconomic status nor the changes that depend on how questionnaires are worded and responses classified for analysis. In the LINZ study, varying the wording in reference to participation and enjoyment in different parts of the questionnaire may have produced the discrepancies reflected in Figure 10.1. Activities which seem to demand a higher degree of commitment, energy and skill than more passive pursuits tend to move up 'popularity tables' when people are asked what they have *enjoyed* and not merely what they have *participated in*.

Sport presents a low profile in the national picture of leisure participation when judged by the small percentage of the adult population who take part in any one particular activity day by day, but the profile is greatly magnified if formal and informal participation in all sporting activities are clustered as one item for analysis, if the focus falls on highlights of the week rather than everyday routines and particularly if participation in sport is taken to include

Table 10.2. Selected leisure activities (by rank) compared across three surveys, New Zealand, 1974–1990.

	1974–75*	1980–81†	1989–90‡
Reading	1	1	1
Television	–	2	2
Listening to music	3	3	4
Visiting friends/family	5	5	3
Gardening	2	7	5
Dining out	11	9	10
Sport	13	8	8
Driving for pleasure	17	11	12

Sources: * Tait (1984), † Department of Statistics (1984), ‡ Cushman *et al.* (1991).

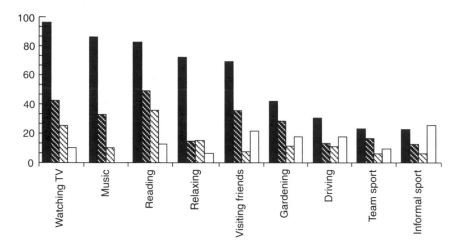

Key: Percentage of respondents listing leisure activities as: (■) participated in; (▨) favourite;
(▩) enjoyed daily; (□) enjoyed weekly
Source: Laidler and Cushman, 1993

Fig. 10.1. Leisure participation, New Zealand, 1991.

spectating, coaching and administration. It then stands out as a majority
interest, as visitors to New Zealand might conclude if they listen and look to
the popular local media for an indication.

Whether to cluster or segregate is, at times, a key question for more than
the presenters of the research findings. For example, agencies with a responsibility for the development and promotion of sport have interpretation choices
to make. They can cluster activities and point to the importance of their role
because sport, leisure and fitness are highly valued by New Zealanders. Or
they can argue that physical activity is undervalued, to judge by low participation rates in some sectors, and that much work lies ahead in the fostering of
healthy lifestyles. Similar alternatives are open to those with a vested interest
in the arts: whether to group active with passive; whether to cluster 'high'
with 'popular' under 'entertainment and culture' (Veal, 1994) to strengthen
a case in defence of policy or practice.

TWO SNAPSHOTS

Full details of the findings from the LINZ study are available elsewhere
(Cushman *et al.*, 1991). Two brief illustrations will be offered at this point to
show how 'snapshots' of social groups can be developed from the national
data, throwing some light and some doubt on common perceptions and thus

providing a backdrop against which other research can be set, using other methods and rationales (Laidler and Cushman, 1991).

Young Men (19–24 Years)

Table 10.3 shows a number of the leisure activities more and less favoured by young male New Zealanders and highlights some of the points at which they seem to differ from their female counterparts and from the general population.

The survey confirmed the predictable, that many young men enjoy being on the move: participating in sport, exercising, running, swimming, boating, and driving for pleasure. But much of their 'leisure time' is spent less vigorously, 'relaxing and doing nothing', listening to music, going to the pub and, overall, the patterns of preference change when other variables are considered along with age and gender. With unemployment, for example, home-based and inexpensive activities, including watching television, gardening, exercising at home and visiting friends, gain favour, in contrast with activities where participation is constrained by lack of money, transport and social confidence, such as going out to restaurants, clubs and pubs, taking part in organized sport and water-based recreation.

Motivation behind participation is notoriously difficult to assess by gathering quantitative data from postal questionnaires, but the answers of the young men, when they were asked what they liked about their favourite activities and why they were attracted to new ones, were encouraging for theorists who see intrinsic satisfaction and pleasure as central to the leisure experience. Most respondents were looking for fun, excitement, novelty and/

Table 10.3. Favourite leisure activities (%), men and women aged 19–24, New Zealand, 1990.

	Men (19–24)	Women (19–24)	General population
Watching TV/videos	38	37	42
Listening to music	37	43	32
Formal sport	27	13	15
Reading	25	38	48
Visiting friends	25	43	35
Visiting pubs/clubs	24	17	10
Dining out	13	20	13
Arts and crafts	6	12	16
Visiting shops	6	19	8
Walking	7	21	22
Gardening	7	11	28

Source: *Life in New Zealand Survey*, 1991.

or competition, rather than for activity as compensation for dissatisfaction in other parts of their lives, as therapy or as a health investment.

Women Over 65 Years

Table 10.4, considered with Table 10.3, allows comparisons to be made between the favourite activities of older and younger adults. Both groups show a number of preferences in line with the general population but each also stands out by recording the highest percentages in certain activities. Women aged over 65 years top the comparative ranking in reading, arts and crafts (including knitting and sewing), looking after pets, driving for pleasure and indoor games. About one in four are involved in voluntary 'work' especially around welfare services. Older women are also prominent in several minority activities, such as going to church and visiting galleries.

The survey reinforced the fact that the majority of women over 65 years are mobile, healthy and active, especially before they join the ranks of the 'old old'. They may not be devotees of strenuous sport and exercise and they may not be comfortable or interested in some leisure environments, such as swimming-pools, rivers and beaches, the bush, restaurants and pubs. And, as energy, confidence, physical capacity, independence and resources dwindle, they drop out of activities and are generally unable to replace them with new interests. But many are enthusiastic about 'going out', both for social inter-action, including 'meeting the opposite sex', and for physical activity. This is the group least willing to 'relax and do nothing' in their leisure. Many are keen walkers and a higher percentage of older women include organized sport and

Table 10.4. Favourite leisure activities (%), men, women aged 65+, New Zealand, 1990.

	Women over 65	Men over 65	General population
Reading	69	60	48
Gardening	52	57	28
Watching TV/videos	49	54	42
Visiting friends	40	27	35
Arts and crafts	34	13	16
Walking	33	35	22
Listening to music	31	23	32
Formal sport	12	13	15
Visiting shops	10	8	8
Dining out	6	7	13
Visiting pubs/clubs	1	12	10

Source: *Life in New Zealand Survey*, 1991.

'being competitive' among their favourite activities than women in the 25–64 age range.

The LINZ study found that New Zealanders undertake 40% of their leisure activities alone, only slightly less than in the most common context: 'with friends'. Sometimes this may be necessary, as with reading; sometimes it may be preferred, as in the case of women hoping to find, in their leisure, time and space in which to be away from other people and their demands. But it is probable that many women over 65 years experience loneliness in their leisure, since nearly 40% of them live alone. In this light, it may be unfortunate that older adults adapt most slowly to technological change if, as a result, they miss opportunities to reduce a sense of isolation. According to the survey, women in the oldest age-band seem in every sense to be least at home with computers and only a very small percentage use video-hire shops.

It was noticeable that the oldest groups had the strongest opinions about such matters as how free time should and should not be spent. It is not known whether the seemingly extreme views represented greater clarity and polarity of judgement, pleasure at being consulted or an intensifying appreciation of the preciousness of their own diminishing strengths and opportunities.

UTILIZATION OF THE SURVEY

The full report of the LINZ survey was published in 1991 in six volumes, including an executive summary and a review of the survey protocol (Wilson *et al.*, 1991). Since then it has been used as a database by the primary research team that was retained as the LINZ Activity and Health Research Unit to carry out further analysis. They have produced over 20 Technical Reports enlarging the statistical snapshots of specific groups caught by the LINZ lens, including Maori, older adults, volunteers, youth in sport, facility users and so on.

The data have also been used extensively by the scientific communities in books, articles and conference papers. They feature in the *New Zealand Pocket Digest of Statistics*. They have been used to various ends by Health Boards and the Nutrition Taskforce reporting to the Department of Health; by libraries and other information services; by marketing organizations; national sports bodies; managers in the leisure industry; planners and policy makers in local government; and lobby groups competing for resources at all levels from national to local community.

Finally, the data have been used for promotion and propaganda. After the LINZ survey, for example, the Hillary Commission for Sport, Fitness and Leisure commissioned a report on the economic and social benefits associated with sport and leisure, entitled: *The Business of Sport and Leisure* (Jensen *et al.*, 1992). The report, which was to become an important reference point for the Commission's strategic planning, continued the tendency to link leisure research very closely with health and exercise, by narrowing the focus from 'sport and leisure' in the title to 'sport and physical leisure' in the body of the report. There was also some continuity of authorship: it included the coordinators of the earlier LINZ and *The Cost of Doing Nothing* projects.

A second example of selective use of the data for the purposes of promoting a professional cause has been provided by librarians, who have used LINZ and the demonstrated high ranking of reading among the nation's favourite leisure activities in their political, social and educational arguments against moves to undermine the 'free library service' principle at the local authority level.

Many of the treasures of the large surveys still lie buried, their data underused and under-analysed. Cost has limited exploration, especially when sponsors have claimed ownership of the information and when funding has been unevenly distributed; that is, covering the costs of initial data collection and the creaming off of propaganda-rich first impressions but not covering follow-up. As discussed at a number of points in this chapter, the surveys were tailored to the fashions and influences of their time. To some extent, goals of comparability were sacrificed in the process.

ISSUES AND PROBLEMS

A number of on-going issues and problems can be identified as a result of the New Zealand experience with leisure participation surveys. They include: the question of presentation and interpretation of findings; the relative importance of sport and leisure; and management and comparability questions related to political and financial dependency and continuity. These are discussed in turn below.

Presentation and Interpretation

In a number of ways, the analysis of the LINZ data was typical of the presentation and interpretation of other national surveys discussed by Hantrais and Samuel (1991): an activity profile was quantified inside a time–space frame; general trends in participation were considered alongside socioeconomic trends assumed to be affecting the leisure context; possible explanatory factors and implications for government were suggested, although this appeared to be more straightforward for analysts using economic rather than social criteria (Jensen *et al.*, 1992). The data allowed for spatial comparisons to be made: that is, of New Zealanders in many places at the same time. The data also created a yardstick for some temporal comparisons of the same activity or facility, investigated at different times.

To some extent, the leisure element in the full survey was vulnerable to the traditional threats of being perceived as residual and non-serious. In the planning of the survey it was secondary to what were seen as the primary indicators of the nation's health. In subsequent policy formulation, advisers to politicians concentrated on the 'serious' aspects of the findings: for example evidence of serious commitment to physical leisure that merited encouragement; evidence of the negative impact of mass inactivity on the nation's health bill; and evidence of the commercial significance of leisure spending and a 'physical leisure industry' too weighty not to be taken seriously.

To some extent also, the survey made it easy for a sanitized residue of leisure to show through since no explicit invitation was given to respondents to comment on their 'irrational' recreation or other activities not normally socially sanctioned.

The Naming of Parts

The survey was planned, executed and interpreted over a period in which there were signs that the idea of *leisure* was gaining prominence, not only as a convenient label for certain goods and services and for a particular professional and academic field, but also as a concept central to discussion of a number of social, cultural and political issues. Among these was the question of whether 'out of work' means 'into leisure' for unemployed youth, for the retired, for people with disabilities, or for women engaged in unpaid labour in the home.

One indicator of such growing prominence was provided by the Amendment Act of 1992. This Act changed the key phrase in the title of the Recreation and Sport Act of 1987 to 'Sport, Fitness and Leisure'. The change introduced the word *leisure* while excluding the term *recreation* and reinforcing the importance of sport and fitness. In the lead-up to the change, the pro-sport lobby had been strong and it was not surprising that the new Act elevated sport in a number of ways: by giving it first name status; by defining fitness and leisure in terms that emphasized their overlap into sport; but by leaving sport free to define itself. The Act proclaims that 'sport means sport'. What *was* surprising was that leisure featured in the title and ambit of the Act, while *recreation* was disowned and left to find a home where it could amid the turmoil of changes in local government, in professional alignments and in the user-pays market-place. Recreation had proved difficult for central agencies that preferred to deal with organizations with strong management structures standing unambiguously inside the territory of sport and physical recreation, exercise and vigorous activity, not arts and crafts or passive inactivity.

Substituting leisure for recreation did not seem likely to remove these difficulties. It can be assumed, however, that the advisers to the Minister argued that there was greater political capital to be gained from the new labelling of the portfolio than from the old or from referring in the title to 'sport' alone. In the first place, narrowing leisure in the text of the Act to 'physical leisure' sharpened the focus of the Minister's area of responsibility but also broadened it, by countering suspicion that 'sport' might mean sport at high levels of performance only. Second, maintaining leisure went some way to mollify the criticism from the local authority level that abandoning recreation represented a shift away from recognition of the social and community values of informal 'non-serious' pursuits. And, third, some of the leaders of business who already had ties with sport and influence on government, could see great commercial and economic potential in a 'leisure sector' which, like fitness, was showing clearer indications than recreation of becoming industrialized. Whether this change in nomenclature will result in a more broad-based approach to leisure and lifestyle in future surveys, or a continuing focus on sport and physical recreation remains to be seen.

Political and Financial Dependency

National leisure surveys in New Zealand have been hampered by the same problems of management and comparability as found elsewhere by Hantrais and Samuel (1991). Management of the research projects has typically been impaired by brittleness and unpredictability in financial backing. Key decisions have sometimes arisen from fortuitously spontaneous combustion: as when a persuasive researcher or an influential lobby group break through bureaucratic filters and tap directly into ministerial power.

In New Zealand, with a small population, the intimacy of single-state politics is facilitated by the comparatively frequent meeting of the comparatively powerful in the main channels of communications, business and travel. Using this advantage, a lobby with access to sympathetic and high-ranking politicians can be strong. Even though the prevailing political and professional ideologies have changed over the last two decades, as recreation, sport, fitness and leisure have been edged out of a welfare and service context towards the market-place of the free economy, programmes and research have won approval because they have been compatible with the portfolios of Ministers of high rank. A previous Minister of Recreation and Sport became Prime Minister. More recently a Minister of Sport, Fitness and Leisure has simultaneously held the politically prominent portfolios of Police and Tourism. Both men proclaimed a deep personal interest in sport and health. The second was outspoken in his hopes that research will demonstrate the value of fitness and physical activity to the health of the nation, the benefits of the leisure industry and recreational tourism to the economy and the potential of sport to reduce social threats to law and order.

The illustration can be viewed in several perspectives. In part it shows the vulnerability of decision making to political and bureaucratic whim and some of the weaknesses of large-scale research projects which, because of their cost, are in a direct line of dependency on political favour, as almost invariably they have been in New Zealand. More pertinently, however, it shows how difficult it is likely to be to replicate the research, as the tides of political ideology change, as funding has to be won on new terms and as new researchers enter the scene, enthusiastic to up-date old methods and avoid old errors.

Comparability and Degrees of Magnitude

The chances of being able to compare like with like over time diminish as the scale and complexity of the research increases. It is extremely difficult to find a 'permanent' place for even one leisure indicator among the principal national benchmarks provided by the census and regular household surveys that are presumed to be politically neutral. The task of finding support for the repetition of large-scale projects like the LINZ survey is at least as difficult, especially if the original has been subjected to the usual accusations: of being open to political and professional manipulation; of being too expensive; or of being flawed in structure and content.

The criticism accentuates the advantages of research that has been planned with reference to internationally agreed criteria and approved practice that runs, as far as possible, on supra-political lines. International cooperation might help identify the key features of the common ground to be explored in national leisure participation surveys. It might also lead to agreement about the measures and indicators to be used internationally and establish principles for systematic analysis as a basis for explanation and generalization rather than for the justification of pre-determined policy.

Initial regional interaction, as between Australia and New Zealand, or among European countries, might build towards comparative work on a wider scale with encouragement from an umbrella organization such as the World Leisure and Recreation Association, the United Nations or a collective of non-governmental agencies like that set up in Geneva to coordinate international collaboration and research related to conservation issues (International Union for the Conservation of Nature and Natural Resources (IUCN)). The gains would be considerable, particularly if comparative frameworks could be set up and research objectives jointly agreed in advance of the data collection.

A sensitive balance will be necessary if the local and the international perspectives are to be complementary. The latter, obviously, can reveal global and historical trends and processes that help explain and even predict local phenomena. In recent years it might have been possible to anticipate the worldwide waves of political, economic and social change that profoundly affected leisure opportunities and attitudes in New Zealand: market forces, conservation, the third age, new technologies, and welfare and local authority services in retreat.

At the same time, however, it will be important to take uniquely local conditions into account when they have a strong bearing on the phenomena being compared internationally. In New Zealand, for example, global questions of how to 'save daylight' and how to deal with an oil crisis (by introducing compulsory 'carless days') were answered idiosyncratically, in ways that profoundly affected leisure patterns. But these measures took place in New Zealand only; they would probably not have been thrown up by comparative surveys that detect enduring structures more effectively than they record 'catastrophes'.

Geographical and demographic factors also dog comparative studies. There are, for example, problems of scale. Compared with Australians, for instance, New Zealanders are in a number of senses easier to survey: they are easier to find and fewer to count. It will be apparent from the above discussion of the LINZ study that many of the researchers' tasks were simplified by advantages such as geographical compactness and centralized communication systems: the vast majority of respondents could be reminded to return their questionnaires by transmitting the message on only one television channel. On the other hand it has also been implied above that small may not be beautiful if one suspects it offers short-cuts to oligarchy. In the research context too, warning bells may ring earlier than in countries where populations are larger. Even a large percentage of a sub-group, such as rural Maori women, may produce a survey cell too small to be statistically useful; the researcher may also discover that survey resistance inside such a sub-group is high as a result of over-exposure to research processes in the past. Resistance might have been amplified by a sense that the processes were culturally intrusive and exploitative. This last warning assumes major importance if attempts are to be made, across international as well as cultural gaps, to standardize methods and train data gatherers. As often, the adequacy of research that draws solely on quantitative data is called into question.

TRACKS TO THE FUTURE

If there are to be further large-scale leisure surveys in New Zealand in the future, the increasing demand for bi-cultural approaches to their design and interpretation will be just one of a number of factors complicating comparative research internationally. It will be possible, no doubt, to identify similar forces advancing certain values, experiences and policy priorities ahead of others. In New Zealand, as elsewhere, there has been a tendency for male, white, middle-class priorities to hold sway and for this to promote the view that the ideal leisure is to be found in physical recreation and sport and the ideal leisure policy is that which encourages the greatest participation in such activity, as a counter to problems threatening the nation's physical, social and economic health (Cushman and Laidler, 1988). It will be tempting to use such a rationale again in the future to win political and financial support, especially if, in all the countries planning joint research, there is sympathy in high places for that rationale.

There are, however, several obstacles to be borne in mind. First, although the mainstream of international trends may carry leisure research in the same general direction, the local cross-currents will always have their effect. They will always be critical in determining whether or not a project is launched and whether it progresses. Second, there will always be pressures on national surveyors to shape the design and results of their research to validate the manifestos and strategies of interested parties in the political, professional and commercial world. Here, the interest will be less in discovering what is actual than in confirming what is deemed desirable. The pressures are unlikely, moreover, to be applied in the best interests of equity.

Widening the research to international scope has the potential to reduce local contamination, but it will continue to be important to examine leisure not as an isolated phenomenon but more broadly, in its social, historical, cultural and economic context; and at the same time more narrowly, in the context of behavioural trade-offs and 'negotiation processes'. This means seeing leisure as more than choice among seemingly unlimited discretionary opportunities (Zuzanek and Larsen, 1993).

REFERENCES

Best, E. (1976) *Games and Pastimes of the Maori*, revised edition. Government Printer, Wellington.

Cushman, G. (1993) New Zealand leisure participation surveys: a preliminary review. In: Veal, A.J. and Weiler, B. (eds) *ANZALS Leisure Research Series*, Vol. 1. Australian and New Zealand Association for Leisure Studies, Sydney, pp. 15–34.

Cushman, G. and Laidler, A. (1988) Recreation and leisure. In: *The April Report of the Royal Commission on Social Policy*, Volume IV. Government Printer, Wellington; pp. 507–531.

Cushman, G., Laidler, A. *et al.* (1991) *Life in New Zealand Commission, Report Volume IV: Leisure*. University of Otago, Dunedin, NZ.

Department of Statistics (1984) *Social Indicator Survey, 1980–81*. Government Printer, Wellington.

Hantrais, L. and Samuel, N. (1991) The state of the art in comparative studies in leisure, *Loisir et Société* 14(2), 381–398.

Hopkins, W., Wilson, N. and Russell, D. (1991) Validation of the physical activity instrument for the Life in New Zealand Survey. *American Journal of Epidemiology* 133, 73–82.

Jensen, B., Sullivan, C., Wilson, N., Berkeley, M. and Russell, D. (1992) *The Business of Sport and Leisure*. Report to the Hillary Commission for Sport, Fitness and Leisure, Wellington.

Jorgensen, M. (1974) *Recreation and Leisure. A bibliography and review of the New Zealand literature*. Ministry of Works and Development, Wellington.

Kelly, J.R. (1990) *Leisure*, 2nd edn. Prentice-Hall, Englewood Cliffs, NJ.

Laidler, A. and Cushman, G. (1991) Life and leisure in New Zealand. In: Veal, A.J. *et al.* (eds) *Leisure and Tourism: Social and Environmental Change*. University of Technology, Sydney/World Leisure and Recreation Association, Sydney, pp. 342–349.

Laidler, A. and Cushman, G. (1993) Leisure participation in New Zealand. In: Perkins, H. and Cushman, G. (eds) *Leisure, Recreation and Tourism*. Longman Paul, Auckland, pp. 1–15.

Perkins, H. and Gidlow, R. (1991) Leisure research in New Zealand: patterns, problems and prospects. *Leisure Studies* 10(2), 93–104.

Robb, M. and Howorth, H. (1977) *New Zealand Recreation Survey: Preliminary Report*. Council for Recreation & Sport, Wellington, NZ.

Russell, D.G., Worsley, F.A. and Wilson, N.C. (1987) *The Cost of Doing Nothing*. University of Otago, Dunedin.

Tait, D. (1984) *New Zealand Recreation Survey, 1974–75*. NZ Council for Recreation & Sport, Wellington.

Veal, A.J. (1984) Leisure in England and Wales. *Leisure Studies* 3(2), 221–229.

Wilson, N., Russell, D., Spears, G. and Herbison, G. (1991) *Life in New Zealand Commission Report Volume II: Survey Protocol*. University of Otago, Dunedin.

Zuzanek, J. and Larson R. (1993) Instead of an introduction: harbingers or revisionism? *Loisir et Société* 16(1), 13–14.

11 Poland

BOHDAN JUNG

LEISURE STATISTICS DURING THE PERIOD OF REFORMS IN POLAND

In the post-Second World War period the bulk of knowledge on leisure participation in Poland came from official statistics routinely gathered by state-owned institutions and firms. The data were then submitted to the Central Statistical Office (GUS) which processed and published them. This process was financed and organized by the state, and resulted in a situation which, in many respects, was extremely comfortable for leisure scholars. Huge statistical studies were carried out at regular intervals, providing a body of data to be analysed and compared. These studies included: time-budget surveys in 1969, 1976 and 1984, carried out on a sample of over 49,000 representative households; annual household budget surveys (with similar sample size); cultural participation studies, involving data from both institutions and population surveys (in 1969, 1971, 1973, 1975, 1982 and 1989, with samples of some 25,000); surveys of participation in holidays and tourism (sequence similar to cultural participation studies above); and annual statistics on culture, sports, recreation and tourism, gathered and published in the country's statistical yearbook.

After the breakdown of the communist system in Poland in 1989, a number of austerity programmes were instituted to reduce the deficit in the government budget. Generally state statistics on leisure participation, with a few exceptions as described below, did not suffer from these cutbacks. The most notable casualty to date has been the time-budget survey, originally scheduled for 1992. Another negative development relevant to the study of leisure participation was a simplification of border traffic statistics which followed liberalization of passport and visa formalities, both for Poles travelling outside of the country and foreigners entering Poland. The previous system of embarkation cards that specified nationality, destination, duration, means of transport and type of stay (business, tourism etc.) was processed by the Ministry of the Interior, which then provided GUS with a detailed breakdown of data. After 1989 this system was replaced by a head count performed reluctantly at border crossings. The gap left by inadequate statistics of border crossings is not adequately filled by occasional surveys carried out by the Institute of Tourism.

Statistics from large-scale studies on leisure participation in Poland in 1991–92 are still available and these form the basis of the discussion in this chapter. Other complementary data originate from smaller, more specialized, studies carried out by such institutions as the Institute of Culture (operated and financed by the Ministry of Culture), the Institute of Philosophy and Sociology of the Polish Academy of Sciences, international projects carried out in Poland as part of assistance programmes (e.g. the PHARE project or *Baseline Study of Polish Industry* quoted in this chapter), or East–West academic initiatives (e.g. the Economic and Social Research Council's project 'Young people and economic change'). To a much lesser extent, data are provided by public opinion polls commissioned by advertising agencies and newspapers. The latter two sources are new to leisure research in Poland. It may be suspected that there is a bulk of knowledge gathered on commercially vital aspects of leisure participation in the country by consultancy firms and market analysts; this knowledge has not, however, been made publicly available and one may only speculate as to its likely impact on evidence on leisure participation.

A breakdown of leisure statistics gathered by the Central Statistical Office (GUS) and other institutions with a short profile of their scope is presented in Table 11.1. The list of statistical sources in Table 11.1, the most recent of which is used in this chapter, is not exhaustive of leisure statistics recently gathered in Poland; it is, however, representative of the mixture of traditional and new sources of information. It also portrays the type of institutions and projects involved in gathering information on current leisure trends in Poland.

OLD AND NEW TRENDS IN LEISURE PARTICIPATION

The evolution of trends in leisure participation can be traced using statistics from national surveys on cultural participation. These data, collected over the

Table 11.1. Sources of leisure statistics in Poland.

Study	Sample size	Representative	Last completed	Survey institution
Time-budget study	49,995	Yes	1984	GUS
Household budget	29,136	Yes	1991	GUS
Participation in culture	11,700	Yes	1990	GUS
Yearbook of culture	–	Complete*	1992	GUS
Participation in tourism	24,000	Yes	1986	GUS
Recreation of the population	69,365	Yes	1984	GUS
Culture and everyday life of Poles	995†	Yes	1991	I.C.‡
Cultural inequalities – Poles/Russians	597	Yes	1988/91	I.C.
Young people and economic change	1,919	Matched sample	1993	ESRC¶
State control of radio and TV	995	Yes	1992	Demoskop§
Do we like going to the cinema?	994	Yes	1993	Demoskop

* Complete statistics for cultural institutions; † 995 households, 2442 interviews; ‡ Institute of Culture; § Public opinion pollster; ¶ (UK) Economic and Social Research Council funding for joint project: University of Liverpool and Warsaw School of Economics.

1972–90 period and presented in a summary form in Table 11.2, allow for generalizations about what can be considered as a relatively new development, and what can be interpreted as a continuation of long-term trends.

Of the 21 activities listed in Table 11.2, five show no clear pattern of evolution, but 16 can be divided into three groups as follows.

1. Those for which participation was lowest in 1972 and highest in 1990. This group consists of only three activities (fishing and hunting, listening to the radio and watching television), with the latter two involving the highest levels of participation in any leisure activity. While the doubling in popularity of fishing and hunting is difficult to attribute to any particular cause, the electronic media group shows all signs of steady long-term expansion.

2. Those with highest participation levels in 1972 and lowest in 1990. The four activities involved – going to the movies, theatre, opera and operetta – exhibit a long-term loss in popularity of those cultural institutions which are representative of a more traditional model of cultural participation. The most notable drop in this group is going to the movies, from over a half of the population in 1972 to under one-third by 1990.[1]

3. Those activities which reached highest participation levels in the mid or late 1980s and experienced a sharp drop in participation (to their lowest or near lowest levels) in 1990. This group is by far the largest and consists of nine activities, including both traditional forms associated with high culture (such as going to museums, exhibitions and concert halls),[2] as well as popular culture pursuits, such as going to shows, amusement parks and circuses. The rapid decline in participation rates over a relatively short period of time could be symptomatic of the short-term effects of economic and political reforms being experienced in Poland at the end of the 1980s.

While limits to generalizations based on the above range of activities have to be acknowledged, the above three groups characterized by clear patterns point to four changes in leisure participation. First, there has been a long-term steady progression in the penetration of households by radio and television, even though growth in participation rates was not spectacular due to relatively high levels of media penetration at the outset of comparative studies in 1972. Second, there has been a long-term decline in traditional forms of cultural participation, with the implied hypothesis that their position in

Table 11.2. Trends in leisure participation in Poland, 1972–1990.

| | Percentage of persons participating in activity | | | | |
	1972	1979	1985	1988	1990
Reading newspapers	88.5	91.6	92.6	87.5	85.9
Reading magazines	68.3	78.0	79.3	74.6	58.0
Reading specialized periodicals	38.7	39.3	42.5	47.4	23.0
Reading books	60.5	46.1	59.8	64.6	57.0
Listening to radio	88.5	89.3	91.0	n.a.	93.9
Watching TV	87.0	97.5	98.4	n.a.	98.7
Going to cinema	53.1	41.1	47.1	n.a.	31.4
Going to theatre	26.3	20.6	14.8	n.a.	13.8
Going to operetta	15.0	13.8	12.9	9.6	8.1
Going to opera	8.6	6.8	6.3	5.7	3.6
Gong to concert hall	5.5	5.3	5.0	9.1	4.5
Going to museum	19.5	14.9	13.0	26.0	11.4
Going to exhibitions	9.1	10.8	8.4	18.2	7.6
Going to circus	28.1	32.0	32.7	29.7	27.2
Going to amusement parks	18.6	22.4	22.1	29.6	19.9
Going to shows	28.1	31.5	33.5	34.5	29.7
Attending sports events	25.3	22.9	25.6	n.a.	23.8
Taking photographs	1.6	8.1	5.3	n.a.	7.1
Doing artwork	4.7	2.9	4.1	n.a.	3.7
Fishing and hunting	5.1	9.4	9.3	n.a.	13.6
Do-it-yourself	1.3	22.7	14.1	n.a.	19.5

Source: *Uczestnictwo w kulturze. Podstawowe wyniki badan reprezentacyjnych z lat 1972, 1979, 1985, 1988, 1990* [Cultural participation. Key findings from representative studies in 1972, 1979, 1985,1988 and 1990], GUS, Warszawa 1992.

people's lives has been replaced by the electronic mass media. Third, the trends reflect the short-term effects of the collapse of those forms of both high and low culture that fared well under a planned economy, state subsidies and little competition from foreign producers. Fourth, there was a move away from collective and sociable forms of participation towards more home based and privatized forms of leisure 'consumption'. Seen from the perspective of McLuhan's 'global village', the range of activities in decline roughly corresponds to those forms of traditional culture in which individualized, local, national or culture-specific contents tend to play a relatively more important role. Correspondingly, the growth in participation in electronic media and their expansionist character, which seems to replace, rather than complement, other activities, could be seen as a symptom of Polish leisure becoming more closely integrated with the global village.[3]

The steady increase in the influence of the media on Polish leisure can be further substantiated using other statistics. For instance, analysis of the data from time-budget studies points to the fact that practically all growth in leisure time in Poland has gone to television watching. The time spent on mass media in Poland rose from 120 minutes per day in 1965 to 122 in 1976, then up to 144 minutes by 1984, accounting for 47%, 50%, and then 53% of total free time. In 1965 television accounted for just 58% of time spent on the media but by 1984 the figure was 76% (Jung, 1990). Furthermore, expenditure on leisure, the share of which actually increased in the household budgets of most groups employed in the state sector from 1985 to 1991, grew mainly as a consequence of increased spending on home electronics, the share of which increased from 14 to 25% of all leisure spending (GUS, Warszawa, 1992). However, the most impressive evidence comes from a census of household durables annually run by the Central Statistical Office (GUS). The rate of replacement of old equipment and purchase of new leisure-related consumer durables increased greatly in the early 1990s. It may be supposed that a heavy increase in the usage of home electronics equipment accompanies any rapid growth in purchases. Similar evidence, pointing to very good material infrastructure for media 'consumption' among Polish youth, was also shown by the 1993 ESRC study *Young People and Economic Change in Poland*.

The move towards more media influence on mass culture in Poland is also confirmed by recent public opinion polls on television usage (*Gazeta Wyborcza*, 10–11.7.93, p. 1). Conducted on a large representative sample of the population aged over 10 years (28,878 respondents), this study showed that the average Pole spends 3 hours 4 minutes per working day in front of a television set. On Saturdays, this time increases to 4 hours 7 minutes and to 4 hours 45 minutes on Sundays. In the course of a lifetime, this corresponds to 10 years of non-stop television viewing, or 15 years if 8 hours sleep is allowed.

While most of the programming for Polish television audiences comes from the two state-owned broadcasting stations (90% and 60% of viewers), some 10–20% watch private or local stations. The same source provides most recent statistics on ownership of electronic media: only 1% of Polish households admit having no television set, 77% watch television in colour, 45% own a VCR, and 20% watch satellite television. Some 88% of the sample watched television every day, 12% watched satellite television at least once a week while 21% watched video at least once a week. The ESRC study *Young People and Economic Change* already mentioned has shown that in the 2 weeks before the survey 63% of young people in two age groups (18–19 and 22–23) watched MTV or video clips, 81% watched commercials on television, and 56% watched foreign (satellite or cable) television.

Another area in which integration of Polish leisure with global trends has become more spectacular in the late 1980s and early 1990s is the movement of people across Polish borders – both Poles travelling abroad and foreigners entering the country. The huge growth in this movement has been evidenced by border statistics. Over the post-war period, the volume of arrivals was largely conditioned by political factors. The introduction of martial law in 1981 brought visitor arrivals down from just over 10 million in 1978 to 1.4 million in 1982. Following the liberalization of the political situation in Poland in 1989 and the formation of the first non-communist government, the number of visitor arrivals in 1990 rose to 18 million, more than double the 1989 level (8 million). In 1991 the number of visitors coming into Poland doubled again to 36.8 million, then increased by a further 50% to nearly 50 million visitors in 1992. The peak year for travel abroad by Poles was reached in 1990, when over 22 million of them visited foreign countries.

The discontinuity in the method of gathering tourism statistics precludes an assessment of precisely how political changes affected Poles' travel to the West. Information appearing in the Polish press, and from pilot surveys, suggests that the Czech and Slovak republics and territories of the former Soviet Union are still the destination of over two-thirds of these departures, with Germany ranking third. In the context of the 'global village', it is also worth noting that liberalization of passport policies was initiated by the communist government nearly 2 years before their being ousted from power and that, over the 1990–93 period, Poles gained the right to enter most of the European Community's member states without visa formalities.

CONTEXT AND BACKGROUND OF CHANGES IN POLISH LEISURE PATTERNS

Mounting evidence on the accelerated integration with the 'global village', which accompanied political and economic changes in Poland at the end of the 1980s, could be regarded as yet further proof of the vitality and attractiveness of mass culture had it not been, as suggested above, replacing, rather than complementing other forms of leisure activities, which are in decline both in terms of participation and spending. This applies not only to high culture, but also to the recreation and sports activities and vacation patterns of the population, which are directly related to fitness and health, already a weak aspect of the country's development.[4] Withdrawal of heavy subsidies for vacations, which had been provided within the framework of the country's social policy (Szubert, 1973), caused massive changes in holiday patterns. Such subsidies had included financial grants for holiday travel, subsistence for schoolchildren and youth and meeting the full cost of operating the so-called 'holiday centres', financed by enterprises, trade unions and professional, government, social and political organizations. These changes led to a lower proportion of people going away from home for holidays and, for those who did, shorter lengths of stay and a higher proportion involving visiting relatives. Sending only children on holidays has become a common strategy of Polish families.[5]

The expansion in the use of electronic mass media as an alternative to all other forms of leisure can also be seen against the backcloth of the general economic situation in Poland. In the course of the economic reforms, the purchasing power of the average household fell by some 40%. In a 3-year period (1990–93) unemployment, non-existent in the communist era, shot up to 16% of the working population (some 3 million, heavily concentrated in geographic terms and among young age groups). In 1989, when the state control over prices was dismantled, the country passed through a wave of hyper-inflation (800%). Throughout the first 3 years of the 1990s, the annual rate of inflation did not fall below 35%. Real interest rates on savings did not keep up with inflation, further aggravating the financial situation of wage earners. Austerity measures resulted in curtailment of the social security system, including cutbacks in retirement pensions, education, health and, of course, culture and recreation services. Seen in this context, Poland's move towards the 'global village' may be interpreted as part of the consumers' strategy to protect their savings from inflation by investing in durables and their determination to protect what they considered to be 'core leisure'.

The data on both incoming and outgoing border flows creates the impression of an enormous tourist boom in Poland, but in fact the record levels of visitor arrivals in 1990 and in 1991 can be attributed mainly to arrivals from Eastern Europe. More detailed statistics by nationality available for 1990 showed that, of the total 18 million visitors, almost 16 million came from Eastern Europe and the Soviet Union, of which some 9 million were from the USSR. According to the definition of 'tourist' by the World Tourism Organization, the majority of these visitors do not qualify as tourists since their stay did not involve an overnight stay in Poland. Visitors from Eastern Europe were mainly crossing borders for a day's shopping and trading. Not using any of the facilities traditionally associated with the tourist industry (accommodation, catering, sightseeing, entertainment), they contributed virtually nothing to tourist revenues (Tourist Development International Consultants, 1991).

The number of visitor arrivals from Western countries, even though not comparable with the volume of visits from Eastern Europe, increased markedly. While in 1981 there were 575,000 visitor arrivals from Western countries, in 1990 the figure reached 2.4 million. While simplification of tourism statistics does not allow for precise assessment of Western arrivals in 1991 and 1992, surveys carried out on small samples (*Życie Warszawy*) suggest that the arrivals of Western visitors after the record year of 1990 are falling off, if not dropping. However, since German visitors from border areas of what was previously East Germany now count as Western visitors, in nominal terms visitor arrivals from the West are booming (in 1990 there were 4.3 million border crossings from East Germany alone, mainly for shopping purposes). Even though these rarely involve overnight stays, they bring in hard currency and have a significant impact on development of the Western border areas of Poland. Surveys conducted by the Institute of Tourism in 1992 show that over 50% of all foreigners entering Poland quoted shopping as their main motivation. Under one-third came on business and only 15% gave tourism and recreation as their objective (*Życie Warszawy*, 31.7.93, p. viii). Analysis of Poles' motivations for going abroad in the late 1980s and early 1990s would also lead to similar conclusions: most of the travel directed to the neighbouring countries involved the so-called 'suitcase importers' and petty traders, not acknowledging their presence in the 'global village' or post-modern universe by seeing the world with what John Urry calls a 'tourist gaze' (Urry, 1990).

A further context in which recent changes in Polish leisure patterns can be evaluated is in relation to statistics reflecting not so much participation, but the supply of leisure goods and services. This allows us to check the results of participation surveys, where participation is declared by respondents (perhaps reflecting imprecise recollections, or projections about what their participation should be) with data on output and attendance gathered by institutions which act as suppliers of cultural goods or services. Such data are available for 1991 for institutions associated with the traditional model of cultural participation, such as libraries and museums, which are mainly public, as shown in Table 11.3.

These data confirm long- and short-term trends discernible from leisure participation surveys. However, the picture that emerges from the producers' point of view seems to be even more dramatic, particularly with respect to falling readership and the crisis of traditional institutions of high culture, including theatres, concert halls and museums, not to mention the cinema, which, under censorship and careful screening and selection of imports in the communist period, had enjoyed the status of a 'cultural event'. Data for radio

Table 11.3. Output and attendance at some cultural activities, Poland, 1980–1991.

Type of activity	1980	1985	1990	1991
Books and pamphlets published (millions)	147.1	246.3	175.6	125.5
Newspapers (total circulation, millions)	2.6	2.5	1.4	1.5
Magazines (total circulation, millions)	0.9	1.0	0.7	0.8
Users of public libraries per 1000 population	208	202	195	178
Borrowings from public libraries per 1000 population	4139	4043	4066	3711
Visits to museums (millions)	20.1	19.9	19.3	15.8
Attendance at theatres/concert halls per 1000 population	500	458	338	279
Cinema attendances (millions)	97.5	107.1	32.8	20.9
Radio listeners (registered users per 1000 population)	243	270	287	281
TV viewers (registered users per 1000 population)	223	254	260	256

Source: 'Kultura 1992', GUS, Warszawa 1992, pp. 2–5.

and television takes into account only registered users (i.e. households paying television and radio licence fees required by Polish law), which explains seemingly lesser penetration levels, but confirms growth in the overall number of users, followed by a slight decline in the 1990–91 period, possibly attributable to an increase in 'wild' viewing, that is, failing to pay licence fees.

In view of the accelerated decline in the number of users/visitors for many cultural activities after 1985, the attempt to typify Polish leisure patterns, undertaken in 1990 on a representative sample of 597, as part of a comparative study of cultural participation of Poles and Russians (Gawda, 1991), seems relatively futile. Using a cluster analysis, the authors of this study established six criteria applied to establish a typology of Polish leisure consumers. Their sample was split into the four groups in the following way:

1. Heavy readers — 23 persons – 3%
2. Television and radio fans — 86 persons – 14%
3. Active participants of all leisure forms — 30 persons – 5%
4. Passive audience — 457 persons – 76%

This typology did not allow for further breakdown of over three-quarters of the sample, classifying it simply as passive (actually meaning residual). Two of the designated groups were not very statistically significant. It can be expected that, with falling levels of participation, even in relatively élitist pursuits associated with high commitment, the share of those difficult to classify is likely to increase in future, making the task of establishing the profile of Polish 'global villagers' even more difficult.

TOWARDS A MORE PERIPHERAL TYPE OF 'GLOBAL VILLAGE' IN POLAND

The transition to a market economy and democratic reforms, which should be viewed as the main context behind changes in leisure patterns in Poland, has resulted in a completely changed situation in the supply of leisure. The predominant character of the Polish leisure market, previously characterized by shortages, institutionalized supply by state firms, subsidized and controlled prices, control over foreign imports, ideological control and censorship in the media, has been completely altered. Accordingly, the traditional issues in Polish leisure research were swept away by new ones. These new issues include the problem of commercialization of leisure provision and its privatization, a growing stratification of leisure consumers by income rather than education due to a rapid polarization of wealth, and lack of economic and personal security in the face of unemployment, soaring crime rates and the appearance of new forms of organized crime, hitherto unknown in Poland.[6]

However, the fundamental problem with Polish leisure can be seen as a cultural one. It has to do with the openness and vulnerability of Polish culture in a new political and economic situation, particularly its ability to survive the massive impact of world mass culture. Unlike most East European countries, in the post-war period Poland has been much more open to the outside (especially the Western) world. This was the consequence of a significant level of emigration to Western countries, resulting in family visits and contacts, more liberal visa and passport policies than other eastern bloc countries, as well as general receptiveness of Polish high culture, which was engaged in a constant dialogue with the West. Under these conditions, it may seem paradoxical to argue that democratic reforms made the country's culture more provincial and peripheral.

To push this paradox even further, the provincialism in question has expanded at a time of increased contacts with the outside world, which were related to the increased flow of people and goods across the country's borders. As mentioned already, the data on spectacular growth in border crossings may create an impression of a tourism boom, while in fact the impact of the increased flows is above all commercial (German shoppers, ex-Soviets as street peddlers, Poles shopping in Germany and in the Czech Republic). The argument on limited cultural impact of a huge increase in border crossings does not hold true for trade flows and virtual lack of customs protection of local industry immediately after the downfall of communism. Huge increases in imports of Western (or rather mainly South-East Asian) leisure durables were accompanied by a spectacular decline in their domestic production (with the exception of VCRs assembled from foreign components), as can be seen in Table 11.4.

Table 11.4. Production of some cultural goods in Poland: 1980–1991.

Type of product ('000s)	1980	1985	1990	1991
Radio and stereo sets	2.7	2.7	1.4	0.6
Colour TV sets	147	158	338	304
Tape recorders	806	384	299	108
Turntables	370	242	127	52
VCRs	–	–	2,014	6,172
Records	880	10,122	4,295	1,825
Cameras	83	43	–	–
Photographic paper (m²)	6,190	6,539	1,995	1,104

Source: 'Kultura 1992', GUS, Warszawa 1992, p. 23.

The true source of provincialism may be attributed to the change in the ethos of the Polish society which accompanied the downfall of communism. During the 45 years of communist rule, there existed little opportunity for creating individual fortunes in Poland. Those who nevertheless succeeded in becoming rich had every reason to hide their wealth. Since material success in a fairly egalitarian society was not feasible, education became the main means of social promotion. It also became the only lasting type of 'human capital' that could not be taxed, nationalized or confiscated. The strong position and ethos of the pre-war Polish intelligentsia added another important element of social advancement and stratification in the form of *cultural capital*, as defined by Bourdieu (1984, 1987).[7] The over-valuation of education and cultural capital, together with blocked opportunities for material success and high consumption levels, created a specific market for high culture, the ostentatious consumption of which was not only fashionable but also acted as a prime identifier of social position and status. By providing democratic access to education and subsidized culture, communism in Poland in fact helped to preserve the nineteenth-century type of literary culture, including such features as the 'cafe society'. Despite ideological control, it also provided a good quality high culture to meet developing demands and fashions. The high standards of this culture trickled down to the media, thus setting the standards for mass culture as well.

With the advent of democratic and economic reforms in Poland, four things happened at the same time. First, there were massive imports of products from the lower end of Western culture (both in terms of hardware and software, accompanied by strong advertising and promotional campaigns). Second, the social imagination was captured by the making of fantastic fortunes (usually illegally) by people who were by no means recognized by the Polish intelligentsia as sharing the same system of values and tastes. Third, there was the downfall of the Polish economy, including the collapse of the public sector and public finance. Finally there was an expansion of individualistic liberalism, fuelled by monetarist and liberal philosophies of market reform which fitted well with the trends then prevailing in international financial institutions.

The ethos of the Polish intelligentsia has been rapidly replaced with materialism, consumerism, ostentation and *nouveau-richism*. The act of watching satellite broadcasts of foreign television on a huge high-definition screen became much more important than *what* was being watched. Given the differences in purchasing power and the lack of cultural competence of the new élites in Poland, participation in culture (and in leisure in general) has become more provincial because the centre-periphery relation has been restored (or, perhaps – made more obvious), in which the lower end of Western mass culture has become much more attractive than local quality high culture.

The problem of the increasingly peripheral and provincial character of Polish culture during the first years of democracy and economic reforms cannot be attributed only to external factors. The change has also come from within. Studies of Polish culture have pointed to cultural chaos beginning to appear even prior to the transition to capitalism (Tarkowska, 1992; Jawlowska *et al.*, 1993),[8] as well as to the unprecedented brutalization and vulgarization of life (Lamentowicz, 1988), which was interpreted by some as aggressiveness associated with entrepreneurship needed in a market economy. Others have seen these phenomena as the cultural heritage of communism, which had a corrupting effect on ethics and norms of social conduct (Tischner, 1992).

The new cultural trends in post-communist Poland will probably facilitate the country's dialogue with postmodernism by introducing a more de-hierarchizing, populist and relativist vision of culture (Featherstone, 1987, 1989; Hall and Jacques, 1989). Such a vision has so far been difficult to accept in the Polish intellectual tradition which favoured heavily structured and normative visions of culture built around élitist high culture and the formation of sophisticated tastes for an avant-garde culture created by the intelligentsia (Milosz, 1989). On the other hand, these trends have successfully alienated many of the country's intellectuals, most of them distinguished dissidents under communism, from the project of creating capitalism in Poland.[9] While too little time has elapsed from the beginning of this project in 1989 to attempt a final judgement, their disenchantment with the way Polish culture is evolving, when seen together with falling trends in nearly all forms of leisure participation (other than the media) are early signs that Poland's accelerated integration with the 'global village' carries a high price, which may not be worth paying.

NOTES

1. This reflects competition from television and video rentals, but also import restrictions on Western films, especially pronounced in the 1980s, when annually only 12–20 new films were shown. Evidence from marketing studies and public opinion polls shows that, with imports of new films and promotion campaigns, the position of the cinema in Poland is showing signs of improvement (*Demoskop*, No.2, 1993).

2. Readership of books, which showed a peak in participation in 1988 and second lowest level in 1990, could also be classified in the same group.

3. However, it is not certain whether this is to be hailed as a sign of (late) modernity, or interpreted as the outcome of the country's weakness and cultural insecurity at a time of transition from one system to another. This idea is elaborated in more detail in the last part of this chapter.

4. In Poland, walking is the only form of physical activity involving participation of more than 10% of the population. There is no single sport which is played by more than 5% of the population (see Jung, 1990; 1992). This is reflected in general health and demographic indicators. According to 1993 data on Poland's demographic situation, despite its relatively young population profile (one-third of its 38.5 million inhabitants are under 20), the average life expectancy (66.7 years for men and 75.7 for women) is among the lowest even by low East European standards, with 52% of deaths being attributed to heart disease (*Gazeta Wyborcza*, 29.12.93, p.1).

5. Conforming to the traditional pattern of the protective family that looks after its children well after their maturity, Polish youth has been relatively shielded from changes in holiday patterns. The evidence from the ESRC's study of young people and economic change in Poland shows that only 23.8% of surveyed youth has not been on holiday away from home; 26.3% has been on holiday once, 20.3% – twice and 29.6% – three or more times. Also, over one-fifth of the sample spent their main holiday abroad. Given their and their parents' living standards and purchasing power, Polish youth is well travelled. In the course of the 2 years before the survey, over 50% visited Germany, over one-third visited another West European country, just under a half went to former Czechoslovakia, over one-third had visited the territory of the former USSR, and one-fifth had been to another East European country. Over two-thirds declare they have been abroad for leisure, tourism or holidays. Only 4% went abroad to study and 21.5% were there to earn money or on business.

6. In the ESRC project, 4.1% of young people in the age groups 18–19 and 22–23 declared they felt threatened in the city centre during daylight, 34.3% felt threatened in the city centre at night and 30.5% felt threatened at night near the place where they live.

7. Cultural capital was defined as the stock of goods which are 'socially' designated as worthy of being sought and possessed (Murdock, 1979, p. 58). The accumulation of cultural capital (an alternative form of accumulation as seen from the perspective of intellectuals and the intelligentsia) leads to an elevation in social status, but it requires cultural competence. This cultural competence consists of three elements: knowledge about the legitimate stock of cultural capital, mastery

of the intellectual and social skills mastering its consumption and use, as well as the ability to deploy this knowledge and skill to advantage of social situation (p. 60).

8. A similar phenomenon of a 'cultural vacuum' has also been observed in other post-communist countries (see Vladimorov, 1992).

9. There are also signs that this disenchantment is not only restricted to intellectuals longing for their social prestige and relatively privileged social position under communism (despite occasional persecution). The sweeping victory of political parties either directly descended from communist parties or representing ideologies opposed to monetarism and liberalism in the democratic elections in October 1993 may be one example. On a lesser scale, the longing for structured cultural order was demonstrated in the 1992 public opinion poll on the need for state control over the programming and contents of what is broadcast or printed in the media, where nearly half of the respondents favoured such control, both in the public and private media (*Demoskop*, No.l, Warszawa, 1993, p. 8).

REFERENCES

Bourdieu, P. (1984) *Distinction: A Social Critique of the Judgement of Taste*. Routledge and Kegan Paul, London.

Bourdieu, P. (1987) The forms of capital. In: Richardson, J.G. (ed.), *Handbook of Theory and Research for the Sociology of Education*. Greenwood Press, New York.

Featherstone, M. (1987) Lifestyle and consumer culture. *Theory, Culture and Society* 4 (1), pp. 55–70.

Featherstone, M. (1989) *City Cultures and Post-modern Lifestyles*. Paper for the 7th European Leisure and Recreation Association Congress, 'Cities for the Future'. Dept. of Admin. & Soc. Studies, Teesside Polytechnic, Middlesbrough.

Gawda, W. (ed.) (1991) *Nierownol'sci w kulturze Polakow i Rosjan* [Cultural inequalities between Poles and Russians]. Instytut Kultury, Warszawa.

Hall, S. and Jacques, M. (eds) (1989) *New Times: The Changing Face of Politics in the 1990s*. Polity Press, London.

Jawlowska, A., Kempny, M. and Tarkowska, E. (eds) (1993) *Kulturowy wymiar przemian spolecznych*. IFIS PAN, Warszawa.

Jung, B. (1990) The impact of the crisis on leisure patterns in Poland. *Leisure Studies* 9(2), 95–106.

Jung, B. (1992) Economic, social and political conditions for enjoyment of leisure in Central and Eastern Europe of 1992 – the Polish perspective. *World Leisure and Recreation*. 34(4), 8–12.

Lamentowicz, W. (1988) Zimny balagan, czyli kulturowy nielad. *Zarzadzanie* 6, 167–169.

Milosz, Cz. (1989) *Zniewolony Umysl*. KAW, Krakow.

Murdock, G. (1979) Class stratification and cultural consumption: some motifs in the work of Pierre Bourdieu. In: Smith, M.A. (ed.) *Leisure and Urban Society*. Leisure Studies Association, London.

Szubert, W. (1973) *Studia z polityki spolecznej* [Studies in social policy]. PWE, Warszawa.

Tarkowska, E. (1992) *Czas w zyciu Polakow. Wyniki badan, hipotezy, impresje* [Time in Poles' life. Results of research, hypotheses, impressions]. IFIS PAN, Warszawa.

Tischner, J. (1992) *Etyka Solidarnosci i Homo Sovieticus* [Ethics of Solidarity and Homo Sovieticus]. Znak, Krakow.

Tourist Development International Consultants (1991) *A Baseline Study of the Polish Tourism Industry.* Dublin-Warsaw.

Urry, J. (1990) *The Tourist Gaze.* Sage, Beverly Hills.

Vladimorov, J. (1992) *Le vide culturel 'postsocialiste' ou les nouvelles orientations culturelles en Europe de l'Est.* EDES, Neuchatel.

STATISTICAL SOURCES

Budzety gospodarstw domowych, GUS, Warszawa 1985, 1988, 1989, 1992 [Household budget surveys].

Demoskop, No. 1, Warszawa, 1993 [Review of public opinion polls].

Demoskop, No. 2, Warszawa, 1993 [Review of public opinion polls].

Kultura 1992, GUS, Warszawa 1992 [Yearbook of cultural statistics].

Maly rocznik statystyczny 1992, GUS, Warszawa 1992 [Concise statistical yearbook].

Rocznik statystyczny 1992, GUS, Warszawa 1992 [Statistical yearbook].

Uczestnictwo w kulturze, GUS, Warszawa 1992 [Cultural participation surveys].

Young people and economic change, Economic and Social Research Council project of University of Liverpool and Warsaw School of Economics, unpublished results of survey.

Zycie Warszawy. Turystyka przyjazdowa i wyjazdowa. *Zycie Warszawy* 31, VII, 1993, p. viii.

12 Spain

CONCEPCIÓN MAIZTEGUI-OÑATE

INTRODUCTION

This chapter has two parts. First, a general overview of research on leisure practices in Spain is presented. Second, the most popular leisure activities that Spaniards participate in are examined using data from secondary sources.

RESEARCH ON LEISURE PARTICIPATION

In Spain, theoretical and research efforts in the leisure field have not run in parallel. During the European Leisure and Recreation Association Congress, held in Bilbao in 1992, Professor Ruiz Olabuenaga mentioned the existing scarcity of empirical research, both in the field of leisure theory and in related areas, such as tourism, sport, and physical education, despite some important developments in theory construction. He nevertheless listed six different areas of leisure-related study represented in the Spanish literature, namely: festivals and collective social celebrations; gambling; sports; tourism; mass media; and leisure and free time as a social phenomenon (Ruiz Olabuenaga, 1992). Leisure education can also be added as an important area of study. During the last quarter of the twentieth century, several Spanish scholars, such as Gines de los Rios, Gil-Albert, Ortega y Gasset, and groups such as the Institución Libre de Enseñanza (Open Education Institution), have been interested in free time and its consequences, with developments at times similar to those of their

European colleagues; in other words, with an important educational or critical component (San Salvador, 1992).

During the Franco dictatorship (1939–1975), leisure scholars and leisure policies were hidden behind 'alternative' movements. For instance, the term 'free time' was always used in an educational context, which its proponents endorsed. Examples are the so-called 'free time clubs' and the scout movement, which have a continuing presence in Cataluña and the Basque Country. These groups not only organized numerous educational activities, they also developed an interesting school of thought and a specific working model. Pedro (1984) argues that free time youth associations have been seen in Cataluña – as in other parts of Spain – as a form of educational intervention focused on specific processes and social development. They are therefore different in character from similar groups in other European countries, such as France. As can be seen in several leisure works (Cuenca, 1984, 1993; Pedro, 1984; Gil Calvo, 1985a; Quintana Cabanas, 1985; Trilla and Puig, 1987; Munne, 1988), academics have also addressed the educational aspects of free time.

However, empirical research aimed at the enhancement of our knowledge of leisure patterns of Spanish society has been very scarce. The majority of available data come from three different sources.

1. Studies from closely related fields, such as tourism, culture, extracurricular activities.
2. Studies focused on particular social groups in which leisure is just one domain of the investigation.
3. General surveys of the population with a section devoted to leisure, but sometimes not even analysed.

In terms of the first source, certain types of leisure activity have received attention from researchers and institutions so that aspects such as sport, cultural activities or tourism can be understood in some depth. Research conducted by García Ferrando (1990, 1993) on sport stands out in terms of the amount of data gathered and its treatment and theoretical context. With regard to one of the most popular leisure activities of the Spanish population, watching television, several ratings studies have been conducted, sponsored by radio and television stations. Televisión Española, a public broadcast service and the only television service available before 1989, conducted this type of study in 1973, 1977 and 1987. Yearbooks of major national newspapers, such as *El País* or *El Mundo* also published results of mass media audience research, giving information on audience segmentation, along with other data. Tourism, an economic sector of great importance in the Spanish economy, has a significant statistical database. However, most of this information focuses on international tourists and their impact on Spanish society and so data regarding the Spanish population's participation in tourism is limited.

A second set of data comes from numerous studies supported by government ministries and foundations (Emakunde, Fundación Santamaría, Ministry of Culture, Ministry of Social Affairs). They have focused on the analysis of particular groups, in particular, youth, the elderly, and women.

Both the Fundación Santamaría and several Ministries have funded projects to understand the social situation and lifestyles of young people (Gonzalo Gallego, 1977; Beltran Villaba *et al.*, 1982; Zarraga, 1984, 1988; Gil Calvo, 1985b; Gonzalez Blasco *et al.*, 1989). Leisure is an important element in all of them. In addition, some regional governments have funded surveys that have yielded detailed information regarding leisure and lifestyles of youth in a number of Autonomous Communities (Elzo, 1986, 1987, 1990). As a result of their methodological rigour and periodicity they allow for statistical comparisons and analyses of the development of leisure patterns over time. Finally, research on leisure and youth has been conducted at a more local level (Calvo and Menendez Vergara, 1985; Aguilera Luna, 1992; Setien, 1992), although sometimes the reports are not readily available outside of the geographical area where they were conducted.

Given the growing demographic importance of the elderly, in recent years there has been an increasing concern with this age group. Some authors have focused on the preventive or therapeutic benefits of leisure for a group whose current life expectancy is now comparatively high (Aguirre, 1981; Castro, 1990; Instituto Nacional de Servicios Sociales, 1990; Subirats, 1990; Mendia, 1991; Martinez Rodriguez, 1992). There are also studies assessing leisure programmes for retired citizens (Instituto Nacional de Servicios Sociales, 1989).

Women have been an important part of the social change which Spain has experienced in recent years, especially in respect to their participation in social life and the work force. Thus, politicians and social organizations have become interested in issues relevant to women. Several studies have sought to explore, from a variety of perspectives, the new social reality of women, including leisure practices and time-budget analyses (Instituto de la Mujer, 1986; Izquierdo and Rodriguez, 1988; Emakunde, 1991; Aretxabala, 1992; Setien, 1992; Salcedo, 1993). In 1983, the Instituto de la Mujer was founded as an institution associated with the Ministry of Culture. One of its main objectives was to seek to understand the situation of women in the legal, educational, cultural, health and sociocultural domains. It therefore includes leisure issues within its terms of reference, particularly in terms of cultural practices.

Finally, the third source of data is general studies at the national level. Some publications, such as those of the Instituto Nacional de Estadística (INE, 1991) have included a section devoted to leisure. Data are presented and compared with results from other countries of the European Community, obtained from secondary sources (e.g. the 'Eurobarometer'). In addition, several universities, sometimes with the support of local governments, have

studied leisure opportunities and/or practices within the immediate commu-
nity, the city or the Autonomous Community (Equipo de Investigación
Interdisciplinar en Ocio, 1991, Cuenca and Setien, 1993 (unpublished);
Ispizua, 1993; Negre, 1993).

Work being done at three institutions that have been the leaders in leisure
research, is worthy of particular mention. They are: Fundación FOESSA, the
Ministry of Culture, and the Centro de Investigación Sobre la Realidad Social
(Research Centre on Social Reality – CIRES). All three organizations, with
different perspectives and historical backgrounds, have regularly conducted
empirical research on leisure participation. The three institutions have notable differences in terms of their evolution, their focus on research topics and
their funding bases. However, they have each conducted research on a regular
basis, at national level, on leisure participation, either from a broad perspective, taking into account a wide range of leisure activities, or from a more
restricted perspective, analysing certain kinds of activity such as culture, sport
or tourism. The reports presented every 5 years by the FOESSA Institute fall in
the first category; the projects developed by the Ministry or by independent
researchers with support from different institutions belong to the second
group. The work of the three institutions is discussed in turn below.

The Fundación FOESSA (FOESSA Foundation)

Beginning in 1965, the Fundación para el Desarrollo de Estudios Sociales y
Sociología Aplicada (Foundation for the Development of Social Studies and
Applied Sociology – FOESSA) published, initially every 5 years, a report on the
social situation in Spain: a series of detailed studies on social change and the
Spanish situation. These are not homogeneous studies, by a single author or
with a specific methodology; they are composite works conducted by several
authors and organized in different chapters. Specifically, beginning in 1975,
they have devoted a chapter to consumption, work and leisure. The leisure
element of the research programme had four goals: first, to establish the
amount of free time available to the Spanish population; second, to discover
and quantify the usage of free time; third, to analyse the social determinants of
the usage of leisure; and fourth, to analyse the different roles of leisure in
Spain.

In 1994, the Foundation was due to publish the results of its latest
research. A vast amount of data on free time was gathered, based on a large set
of questions. The nationally representative sample consisted of 8500 subjects
over 18 years of age. The activities list contained a total of 24 different leisure
practices and data on membership of a variety of types of association (political,
cultural, professional, sports). Professor Ruiz Olabuenaga, of Deusto University, has been responsible for the chapter entitled 'Leisure and lifestyles'.

The main objective of the analysis has been to broaden the scope to include *lifestyles*. In order to do so, leisure typologies have been developed using time and space categories, resulting in seven lifestyles. The study also had two secondary objectives: (i) to establish what free time activities people participate in and their frequency; and (ii) to analyse social life.

Ministerio de Cultura (Ministry of Culture)

The Ministerio de Cultura periodically conducts research on the cultural practices of Spaniards. They have applied a broad perspective including activities that have traditionally been considered culture, as well as activities directly related to leisure and free time – such as visits to amusement parks – and hobbies and other recreational activities that take up people's free time (Ministerio de Cultura, 1991). The latest investigation was presented in 1991, with a sample of 15,000 subjects. In each household selected, only one individual was interviewed so that the respondents represented households as well as individuals. The goals of the project were to discover and quantify the cultural resources of the Spanish families; to establish and quantify the relationship between household members and such resources; to establish and quantify the cultural practices and patterns of the Spanish families; and to gather data on the current cultural patterns of the Spaniards. The survey shows the practices Spaniards prefer, the frequency of participation in each activity and the activities engaged in during the last year. The result is an extensive map of Spanish cultural life, from cultural practices to recreational and educational activities.

Centro de Investigación Sobre la Realidad Social (Research Centre on Social Reality)

The Centro de Investigación Sobre la Realidad Social (CIRES), is funded by three financial institutions: Fundación Bilbao-Vizcaya, Caja Madrid, and Bilbao Vizcaya Kutxa. It regularly publishes monographs on the results of research on various aspects of 'social reality', such as the mass media, participation in public life, time usage, and the elderly. Its work is characterized by its methodological rigour, the interest of the topics addressed, and the free circulation of the data in the scientific community so that secondary analyses can be conducted. Specifically, the issues relevant to leisure are those devoted to time-budgets (time usage) and the monographs on mass media.

Discussion

In Spain, the historical lack of research in every single field of human knowledge has been even more apparent in the leisure domain (Ruiz Olabuenaga, 1992; San Salvador, 1992). Given this situation, it is necessary to emphasize the need for leisure research conducted by leisure scholars. Empirical data alone are not sufficient and thus it is necessary to encourage conditions that can foster much needed theoretical work.

As a result of the lack of a theoretical framework the collection and analysis of empirical data has often been less than ideal. Sometimes survey results have been analysed applying confusing criteria to group the data. Categories have not been clearly defined. Leisure has been considered, either as an all-inclusive construct or as a construct in which leisure domains such as sports or mass media consumption were excluded.

Having a set of empirical data on leisure practices is but a first step to exploring their relative meanings, the role of leisure in social change, the transformation and adaptation of leisure to new social conditions, its personal meaning, its educational possibilities, and its economic and social impact. It is also necessary to acknowledge the difficulties involved in research on leisure participation. A thorough analysis should include all possible leisure practices, so that fields such as culture, sport and tourism can be taken into account. Nowadays general sociological reports tend to add a chapter focusing on leisure participation, for example, *España 1993*, published by the Centros de Estudios del Cambio Social (Centre for the Study of Social Change) (1994), which includes a chapter on the use of free time by Luisa Setien.

Another problem resulting from the research situation described above is the difficulty in conducting longitudinal studies due to the changes in the design of periodical surveys. In this regard, work being done by several researchers is of interest. Using secondary data, Gil Calvo (1985a), Piñuel (1987), Ramos Torres (1990), Zorrilla (1990) and Escobar de la Serna (1991) have studied the meanings of leisure in Spanish society. Escobar de la Serna, Gil Calvo, and Piñuel focused on audio-visual media and their consumption. Ramos conducted a complete analysis of the activities of Spaniards comparing two sets of data gathered in two different research projects – the survey on free time usage conducted by the Social Research Centre for Spanish Television, and the survey on activities conducted by OTR/IS for the Institute for Women. The author uses a typology similar to that used by As (1978), though he does not follow his terminology. The section on free time includes relaxation, recreation, and cultural consumption activities, grouped in three broad areas: shows, amusements and social activities; sports and active leisure; and passive leisure. Since the original data were based on different categorization principles, the author had to overcome some serious problems in comparing the data sets.

Traditionally, Spanish scholars have thought of leisure as the appropriate context for the development of personal and social skills as well as for the development of self-esteem, given that the contribution of leisure to the quality of life and life satisfaction seems obvious (Setien, 1993). However, it seems puzzling that, even though this hypothesis has been accepted at the theoretical and pragmatic level, in the field of empirical research very few studies have addressed the issue. Empirical research on participation in leisure activities is scarce, even more scarce is qualitative research that adds further information rather than just data on frequency of participation. The recently published work by Negre (1993) can be mentioned, as well as the work by Retolaza (1988) and Maiztegui and Retolaza (1989).

Negre focused on participation from a life cycle perspective, using a qualitative methodology involving in-depth interviews. This approach allowed the inclusion of psychological aspects which complemented the information derived from the report of activities. Research by Retolaza has centred on the assessment of educational interventions through free time youth organizations, that is, the measurement of changes in young people's attitudes (social, religious, ecological and cultural) and personal development, resulting from participation in free time groups.

Recently, there has been an interest within certain groups of disabled citizens stemming from the difficulties encountered in their participation in leisure activities. In-depth analysis of the psychological, social or structural barriers and constraints on participation have not, however, been developed. An example of studies in the field of disabilities is the work on mental health by Larrinaga *et al.* (1993), studies on leisure preferences and participation of visually impaired individuals sponsored by the Organización Nacional de Ciegos de España (Spanish National Association for the Blind, Equipo de Investigación Sociológica, 1993), and the studies on the accessibility of tourist sites by the Basque Community Association of Physically Disabled (Siadeco, 1993).

Finally, in Spain, there is a lack of research that, from a quantitative or qualitative perspective, goes beyond the descriptive stage to elaborate a theoretical framework supported by the human and social sciences and which, as Pronovost and D'Amours (1990) argue, might form new ways of questioning and understanding for individuals and society.

MAIN LEISURE PRACTICES OF THE SPANISH POPULATION

This section analyses the major leisure practices of the Spanish population, drawing on data from a number of the research projects presented above. First, the amount of free time available is discussed, and this is followed by an examination of the leisure activities engaged in during that time. Leisure activities have been classified into five categories: family and social relations, mass media, membership of organizations, sporting activities, and cultural activities.

Studies on time-budgets show that in the allocation of free time there are clear differences between men and women. During the week the difference is an hour and a half but at weekends it increases to two and a half hours. Age and marital status (being married) seems to increase the male privileges of free time. Except for the youngest (14–18 years) and the oldest (over 64 years) age groups, the differences in time devoted to household chores by males and females increases during the weekend.

Family and Social Relations

Family life and use of the mass media are the most popular leisure activities among the Spanish population, as shown in Table 12.1. The relationship of these two activities is obvious, given that the majority (97%) watch television with the family. The Spanish family has changed and adapted to new times and, as Gaullier (1988) notes, it has evolved from a 'horizontal' to a 'vertical' form. The trans-generational family has appeared and, for the first time, four different generations may live together. Thus, families have become more complex. The stereotype of solitude for the elderly has been replaced by different kinds of relationship, based on respect and confidence. The family continues to be one of the main reference groups and sharing free time with the family is the major leisure activity. Satisfaction with family life has increased since 1981, both in terms of the percentage of satisfied individuals and the measure of general satisfaction. In general, the family atmosphere appears to be good, without major conflicts. Both the young and the elderly seem satisfied.

For youth, the role of the peer group is also important as the primary reference group with which they spend most of their free time. As Negre (1993) points out, leisure during these years is pure sociability. On the other hand, the elderly feel supported by their families which in turn affects their assessment of their own life. Some 79% report being satisfied with their current life.

Mass Media

In western societies, including Spain, every single individual is a consumer, directly or indirectly, of the mass media. As Zorrilla has pointed out: 'the mass media, particularly television, continue to be the major reference of the post-consumption society' (Zorrilla, 1990, p.19). The most common cultural

Table 12.1. Participation in leisure activities, Spain, 1992.

	% participating*		
	Males	Females	Total
Be with the family	77.3	84.4	81.0
Watch TV	66.9	70.3	68.6
Be with friends	56.1	45.7	50.7
Read magazines and books	20.9	25.8	23.4
Listen to the radio	19.6	28.1	24.0
Listen to music	16.1	15.2	15.7
Outings	14.8	15.3	15.0
Watch sports	17.2	2.5	9.6
Practise sports	13.8	4.6	9.1
Attend movies	6.3	7.0	6.7
Attend shows	2.4	1.9	2.1

Source: FOESSA, 1994.
* No time period specified.

resources are those that serve as vehicles for mass media messages, especially television, radio and the radio-cassette. Some 98% of Spanish households have at least one television set, this being a colour set in 89% of cases. The electronic media are followed at some distance by the so-called 'de-massed mass media', or media which are targeted to a restricted group. It can be said that the mass media reach everybody and, as the Report of the Ministry of Culture concludes, 'the lack of equipment is an almost residual situation, always below two per cent, and can be explained by the critical situation of the household or is directly related to the location of the residence – rural or urban – of the families' (Ministerio de Cultura, 1991, p. 39).

Several investigations (CIRES, INE, Ministerio de Cultura) show that the Spanish population has frequent and direct contact with radio and television. Both have a committed audience that follows their programmes on a daily basis. Among all mass media, television is preferred by all social groups and by people of all ages and levels of education. As Kelly and Godbey have said (1992, p. 455): 'Television is *the* mass medium'. The average time Spaniards devote to television watching is approximately 2 hours a day, although important differences appear. Some 18% report watching television less than 45 minutes daily, and 13% more than 3 hours a day (CIRES, 1992). The key variables seem to be age and socioeconomic status: time spent in television watching increases with age, and decreases with increases in socioeconomic status and size of residence (CIRES, 1992). These results are similar to those found in other European countries. Time spent watching television is an important proportion of all free time available. Within Europe, the British and the Spaniards spend most time in front of the television – a total of 207 minutes a day on average (INE, 1991).

Spaniards prefer informative programmes and films, although half of them report watching television as a secondary activity while they do something else, such as talking (24%), eating (47%), doing household chores (15%) or reading (6%). These patterns seem to be broadly true of all social groups. There is no difference in time spent on television watching for men and women, but women tend to simultaneously perform household chores. As Dumazedier (1989) pointed out, television could be a catalyst for new relationship patterns, and could facilitate family conversations and discussion regarding the programmes viewed. Most of the time (78%), people watch television with someone else, mainly family members, and they report having conversations about what is being viewed.

Other mass media, except for the radio, are not as popular. There is a low preference for reading newspapers – only 49% of the Spanish population reports doing so. Newspaper reading patterns vary greatly among social groups: people over 65 and those with a low socioeconomic status comprise more than the 40% of those who never read the daily press (CIRES, 1992). The majority of newspaper readers are males. In terms of magazines, women seem to be the consumers of the most popular magazines, that is the 'pink' press.

Membership of Organizations

Membership of organizations in Spain does not reach the participation level of other European countries, perhaps due to a certain resistance on the part of Spaniards to associate. In 1993, sports associations had the highest membership, at 14.2%, which is far behind those of countries such as Ireland, Denmark and the Netherlands, which reach a membership level of 30%, but it represents an increase over the 1987 figure, which was only 5.4% (INE, 1991).

Sport

Sport participation shows gender and other social differences: males, young people and individuals with higher educational levels participate more in sport. Marital status also distinguishes sports participation, singles being more involved in sports activities than married people.

Cultural Activities

Use of cultural centres, a facility of recent history in Spain, deserves special attention. Funding organizations have favoured the establishment of sports, cultural and recreation centres and participation has increased in recent years. Visits to monuments and museums have become popular, and cultural facilities have become major tourist facilities. Certain cultural events are enjoying great success in terms of public support, for example the Thyssen Museum, opened in Madrid in 1992 and the recent Velazquez Exhibition, in the Museo del Prado, attracted 700,000 visitors. At the time of writing, the Goya Exhibition in the Museo del Prado, and the Master Works Exhibition of the Guggenheim Collection in Bilbao, appear to be following a similar pattern. During their holidays, 31% of the population report visits to heritage centres, 23% visit museums and 15% visit national parks (considered as cultural heritage by the state). However, according to the Ministry of Culture survey, participation in cultural activities is influenced by social status and educational level.

Geographical location seems to affect the cultural interests of the population. Madrid and Northern Spain stand out for their dynamism and level of cultural interest. Their populations are more acquainted with book shows, music festivals, theatre and cultural visits. This reflects historical tradition stemming from earlier urban and economic development, even though the current economic crisis particularly affects the northern regions, with their economy based on mining and the iron and steel industry. Inhabitants of urban areas seem to prefer cultural trips and reading, though level of education and age also account for the high levels of participation in these activities (Ministerio de Cultura, 1991).

Gender, age and cultural capital are variables known to affect social life and thus participation in cultural activities. As Negre notes:

> Leisure practices are the kaleidoscopic result of multiple lifestyles somehow shared, of preferences and interests acquired within one's environment, group or social class that tend to reproduce, change and reflect from one domain to another.
>
> (Negre, 1993, p. 27)

Discussion

In sum, participation in leisure activities in Spain is not much different from other European countries. However, the important role of family and social relations is worth mentioning. In addition the impact of mass media on leisure, particularly television, has been marked. Its presence in the Spanish household has shown a significant increase. In general, television is the preferred resource for free time. Caffarel Serra (1992) believes that 'this reinforces the entertainment functions of television while the informative functions are primarily fulfilled by the press' (Caffarel Serra, 1992, p. 225).

Based on frequency of participation in leisure activities, the 1993 FOESSA report establishes four types of leisure that evolve into seven types of leisure lifestyles.

1. *Daily home-based leisure*, which includes daily free time activities such as television watching, chatting, playing cards.
2. *Non-daily home-based leisure*, which includes family gatherings and celebrations, anniversaries and the like.
3. *Daily out-of-home leisure*, or those activities performed out of the home but on a daily or regular basis: sports, outings with friends, etc.
4. *Non-daily, out-of-home leisure*, such as vacations and holidays.

The majority of leisure activities have an economic aspect that has made them part of the family budget. In times of crisis, economic difficulties affect the family budget and thus money allocated for leisure is dramatically reduced. The Spanish Association of Consumers (UCE, unpublished data) believes that almost 70% of Spanish households have, in the current recession, reduced their leisure expenses considerably, including expenditure on movies, theatre-going and trips. However, leisure practices are not isolated events. They can be considered as relevant aspects of identity formation in advanced societies (Kelly, 1985), and they permit the formation of personal lifestyles. As Ruiz Olabuenaga (1992) believes, they can be considered lifestyles because they are frequently practised, in a systematic and continuing fashion, and cannot be considered as exceptional behaviour.

CONCLUSIONS

Even though leisure research in Spain does not have a long history, we currently have a fairly complete picture of leisure participation. This information is of interest not only to researchers but also to public bodies responsible for the provision of leisure facilities and services. It is also of interest to the services sector where leisure activities (cultural, touristic or recreational) are one of the fast growing sub-sectors. However, the fact that leisure activities are related to other factors such as marketing and motivation should not be forgotten. These factors need to be better understood in order to comprehend the social situation generally. In this regard, several multi-disciplinary research projects are underway in Spain in order to fully understand the leisure participation phenomenon (Equipo de Investigación Interdisciplinar en Ocio, 1991; Cuenca and Setien, 1993 (unpublished)).

The difficulties encountered in the process of gathering data on research projects conducted in different fields (academic, public sector, private sector) are partly due to the problems inherent in national surveys and thus local research tends to predominate. Besides, university leisure departments are very recent in Spain and thus scholars continue to work within their original fields and to publish in their separate professional journals. There is a great need to develop a leisure forum where research results and ideas can be discussed and shared among interested investigators.

Despite the situation, there is room for optimism. First of all, leisure research is reaching an increasing number of interested specialists from different disciplines. Second, we should not forget our history of theoretical work on leisure and related issues (tourism, sports, culture, and mass media) which offers a solid background to formulate and answer important research questions.

REFERENCES AND FURTHER READING

Aguilera Luna, J.L. (1992) Investigación sobre la situación de los jóvenes de la provincia de Málaga en el campo del ocio y del tiempo libre. Paper presented at the *VIII European Leisure and Recreation Association Congress*, Universidad Deusto, Bilbao, July.

Aguirre, J. (1981) *Ocio Activo y Tercera Edad: Un Proyecto Comunitario*. Caja Laboral, Mondragon.

Aretxabala, E. (1992) Estilos deportivos femeninos a nivel recreativo. Paper presented at the *VIII European Leisure and Recreation Association Congress*, Universidad Deusto, Bilbao, July.

As, S. (1978) Studies of time-use problems and prospects. *Acta Sociologica* 21, 121–141.

Bazo, M.T. (1990) *La Sociedad Anciana*. CIS, Madrid.

Beltrán Villaba *et al.* (1982) *Juventud Española 1960–82*. Fundación Santamaría & S.M, Madrid.

Caffarel Serra, C. (1992) El ocio y MCM. *Revista Española de Investigaciones Sociológicas* 57, 213–226

Calvo, E. and Menendez Vergara, E. (1985) *Ocio y Prácticas Culturales de los Jóvenes*. Publicaciones de Joventud i Societat, Barcelona.

Capel Martinez, R. (1986) *Mujer y Situación Social en España 1700–1975*. Ministerio de Cultura, Madrid.

Castro, A. (1990) *La Tercera Edad: Tiempo de Ocio y Cultura*. Narcea, Madrid.

Centro de Estudios de la Realidad Social (1991) *El Uso del Tiempo*. Boletín CIRES, Febrero, Madrid.

Centro de Estudios de la Realidad Social (1992) *Medios de Comunicación*. Boletin CIRES, Diciembre, Madrid.

Centro de Estudios de la Realidad Social (1993) *Familia y Uso del Tiempo*. Boletín CIRES, Febrero, Madrid.

Centros de Estudios del Cambio Social (1994) *España 1993: Una Interpretacion de su Realidad Social*. Fundación Encuentro, Madrid.

Cuenca, M. (1984) *Educación Para el Ocio*. Cincel, Madrid.

Cuenca, M. (1993) Ocio y futuro: Del homo ludens al homo festus. *Letras de Deusto*, 23, 59.

Dominguez, I. (1991) El Reparto del Tiempo Libre. Paper presented at the *VIII European Leisure and Recreation Association Congress*, Universidad Deusto, Bilbao.

Dumazedier, J. (1989) *Révolution Culturelle du Temps Libre*. Méridiens Klinscksieck, Paris.

Elzo, J. (1986) *Juventud Vasca 1986: Informe Sociológico*. Servicio General de Publicaciones del Gobierno Vasco, Vitoria-Gazteiz.

Elzo, J. (1987) *Una Lectura de la Juventud Vasca*. Servicio Central de Publicaciones del Gobierno Vasco, Vitoria-Gazteiz.

Elzo, J. (1990) *Jóvenes vascos 1990: Informe Sociológico*. Servicio Central de Publicaciones del Gobierno Vasco, Vitoria-Gazteiz.

Emakunde (1991) *Informe Sobre la Situación de las Mujeres en Euskadi*. Emakunde, Vitoria-Gazteiz.

Equipo de Investigación Interdisciplinar en Ocio (1991) *El Ocio en el área Metropolitana de Bilbao*. Ayuntamiento de Bilbao, Bilbao.

Equipo de Investigación Sociológica (1993) *Las Necesidades en Servicios Sociales de los Afiliados a la ONCE*. ONCE, Madrid.

Escobar de la Serna, L. (1991) *La Cultura del Ocio*. Eudema, Madrid.

Fundación FOESSA (1976) *Informe Sociológico Sobre la Situación Social en España*. Euroamérica, Madrid.

Fundación FOESSA (1994) *Informe Sociológico Sobre la Situación Social en España*. Euroamérica, Madrid.

García Ferrando, M. (1990) *Aspectos Sociales del Deporte*. Alianza, Madrid.

García Ferrando, M. (1993) *Tiempo Libre y Actividades Deportivas*. Instituto de la Juventud, Madrid.

Gaullier, X. (1988) *La Deuxième*. Seuil, Paris.

Gete Alonso, E. (1987) *Tiempo de Ocio*. Plaza y Ganés, Barcelona.

Gil-Albert, J. (1982) *El Ocio y sus Mitos*. Imprenta Sur, Málaga.

Gil Calvo, E. (1985a) *Los Depredadores Audiovisuales*. Tecnos, Madrid.

Gil Calvo, E. (1985b) *Ocio y Prácticas Culturales de los Jóvenes*. Ministerio de Cultura, Madrid.

Gonzalez Blasco *et al.* (1989) *Jóvenes Españoles 89*. Fundación Santamaría, Madrid.

Gonzalo Gallego, L. (1977) *Ocio y Vida Cotidiana de la Juventud Trabajadora*. Popular, Madrid.

INE (1991) *Indicadores Sociales*. Instituto Nacional de Estadística, Madrid.

INSERSO (1989) *Ocio en la Tercera Edad*. Ministerio de Asuntos Sociales, Madrid.

Instituto de la Mujer (1986) *Situación Social de la Mujer en España*. Ministerio de Cultura, Madrid.

Instituto Nacional de Servicios Sociales (1989) *El Ocio en la Tercera Edad: Programas de Vacaciones*. Ministerio de Asuntos Sociales, Madrid.

Instituto Nacional de Servicios Sociales (1990) *La Tercera Edad en España: Necesidades y Demandas*. Ministerio de Asuntos Sociales, Madrid.

Ispizua, M. (1993) *Hábitos Deportivos de la Población de la Margen Izquierda*. Diputación Foral de Bizkaia, Eudel, Gobierno Vasco, Bilbao.

Izquierdo, J. and Rodriguez, A. (1988) *La Desigualdad de las Mujeres en el uso del Tiempo*. Ministerio de Asuntos Sociales, Madrid.

Kelly, J. (1985) *Leisure Identities and Interactions*. Unwin Hyman, London.

Kelly, J. and Godbey, G. (1992) *Sociology of Leisure*. Venture, State College, PA.

Larrinaga, V. and Sanchez, J. (1991) Salud Mental, una Experiencia de Intervención Dentro de un Programa de Ocio. Paper presented at the *VIII European Leisure and Recreation Association Congress*, Universidad Deusto, Bilbao, July.

Larrinaga, V. *et al.* (1993) Escala de ocio, Validez, Fiabilidad y Aplicaciones. Paper presented at the *3rd International Therapeutic Recreation Symposium*, Toronto.

Lozares, C. (1991) Vida Cotidiana, Distribución de Tiempos y de ámbitos y Prácticas de Tiempo Libre. Paper presented at the *VIII European Leisure and Recreation Association Congress*, Universidad Deusto, Bilbao, July.

Maiztegui, C. and Retolaza, J.L. (1989) The Role of Free Time Clubs in the New Educational Leisure Policy in Bilbao. Paper presented at the *VII European Leisure and Recreation Association Congress*, Rotterdam.

Martinez Rodriguez, S. (1992) La Preparación para la Jubilación. Hacia un Nuevo Tiempo de Ocio. Paper presented at the *VIII European Leisure and Recreation Association Congress*, Universidad Deusto, Bilbao, July.

Mendia, R. (1991) *Animación Sociocultural de la Vida Diaria en la Tercera Edad*. Documentos de Bienestar Social del Gobierno Vasco. Vitoria-Gazteiz.

Ministerio de Cultura (1978) *Demanda Cultural de los Españoles*. Ministerio de Cultura, Madrid.

Ministerio de Cultura (1985) *Encuesta del Comportamiento Cultural de los Españoles*. Ministerio de Cultura, Madrid.

Ministerio de Cultura (1991) *Encuesta de Equipamiento, Prácticas y Consumos Culturales*. Ministerio de Cultura, Madrid.

Moragas, R. (1991) *Gerontología Social. Envejecimiento y Calidad de Vida*. Herder, Barcelona.

Munne, F. (1988) *Psicosociología del Tiempo Libre. Un Enfoque Crítico*. Trillas, México.

Negre, P. (1993) *El Ocio y las Edades*. Herder, Barcelona.

Pedro (1984) *Ocio y Tiempo Libre. Para qué*. Humanitas, Barcelona.

Piñuel, J.L. (1987) *El Consumo Cultural*. Fundamentos, Madrid.

Pronovost, G. and D'Amours, M. (1990) Les études du loisir: pour une nouvelle lecture de la société. *Loisir et Société* 13(1), 15–38.

Quintana Cabanas (1985) *Fundamentos de Animación Sociocultural*. Narcea, Madrid.

Quintana Cabanas (1993) *Ambitos Profesionales de la Educación*. Narcea, Madrid.

Ramos Torres, R. (1990) *Uso del Tiempo Libre y Desigualdad entre Mujeres y Hombres en España*. Instituto de la Mujer, Ministerio de Asuntos Sociales, Madrid.

Retolaza, J.L. (1988) Evaluación de los resultados del asociacionismo juvenil de tiempo libre. *Letras de Deusto* 18(42), 171–175.

Ruiz Olabuenaga, J.I.(1992) La investigación del ocio en España. *Proceedings of the VIII European Leisure and Recreation Association Congress*, Universidad Deusto, Bilbao, July.

Ruiz Olabuenaga, J.I. (1993) *Ocio y Estilos de Vida en Fundación FOESSA*. Informe FOESSA, Madrid.

Salcedo Miguel, M. (1993) *Participación Femenina en el Deporte*. Emakunde, Vitoria-Gazteiz.

San Salvador, R. (1992) Ocio y pensamiento en España. *Letras de Deusto* 57, 133–146.

Setien, M.L. (1992) Uso del Tiempo y Ocio en las Jóvenes Universitarias vascas. Paper presented at the *VIII European Leisure and Recreation Association Congress,* Universidad Deusto, Bilbao, July.

Setien, M.L. (1993) *Indicadores de Calidad de Vida.* CIS, Madrid.

Siadeco (1993) *Guía de Accesibilidad de Servicios Turísticos de la Comunidad Autónoma del País Vasco.* Confederación Coordinadora de Disminuidos Físicos de la Comunidad Autónoma del País Vasco, Bilbao.

Soler i Maso, P. (1992) El Ocio Como Actitud Educadora. Estudio Desde la Realidad Catalana. Paper presented at the *VIII European Leisure and Recreation Association Congress,* Universidad Deusto, Bilbao, July.

Subirats, J. (1990) *La Vejez Como Oportunidad.* Ministerio de Asuntos Sociales, Madrid.

Trigo, E. (1992) Estrategias de Futuro Para un Ocio Creativo. Paper presented at the *VIII European Leisure and Recreation Association Congress,* Universidad Deusto, Bilbao, July.

Trilla, J. and Puig, J. (1987) *La Pedagogia del Ocio.* Laertes, Barcelona.

Trilla, J. and Puig, J. (1989) *La Integración de Niños y Jóvenes con Discapacidad en Actividades de Tiempo Libre.* Ministerio de Asuntos Sociales, Madrid.

Viché, M. (1986) *Animación Sociocultural y Educación en el Tiempo Libre.* Group Dissabte, Valencia.

Zarraga, J.L. (1984) *El Empleo del Tiempo y Recursos Económicos de los Jóvenes Españoles.* Ministerio de Cultura, Madrid.

Zarraga, J.L. (1988) *Informe Juventud en España.* Ministerio de Asuntos Sociales, Madrid.

Zorrilla, R. (1990) *El Consumo del Ocio.* Consejería de consumo del Gobierno Vasco, Vitoria-Gazteiz.

13 United States of America

H.K. CORDELL, B.L. McDONALD, B. LEWIS, M. MILES,
J. MARTIN AND J. BASON

INTRODUCTION

This chapter is based on the National Survey on Recreation and the Environment, which represents the continuation of the National Recreation Survey (NRS) series, begun in 1960 by the Outdoor Recreation Resources Review Commission (ORRRC) (1962).[1] The first NRS was a four-season, in-home survey of outdoor recreation participation in the United States. Since that time, five additional NRSs have been conducted, in 1965, 1970, 1972, 1977, and 1982–83. Despite the advantages of a continuing series, the NRS has suffered from a lack of continuity in funding, sponsorship, methodology, and composition. For example, the 1965 survey was restricted to post-summer interviews, the 1970 NRS was a short mail supplement to the nation's Fishing and Hunting Survey, and the 1977 survey was conducted by telephone. In addition, different federal agencies have taken responsibility for the NRS throughout the years, further impairing its continuity. These differences have made valid comparisons difficult but still possible, because a core set of participation and demographic questions has been retained.

The ORRRC recommended that a national recreation survey be conducted every five years, however, no specific and consistent funding was identified for this purpose. To complicate matters, the agencies responsible for conducting the NRS themselves experienced a number of transitions over the ensuing years. Between 1965 and 1977 the surveys were conducted by the Bureau of Outdoor Recreation and its successor, the Heritage Conservation

© 1996 CAB INTERNATIONAL. *World Leisure Participation* (eds G. Cushman. A.J. Veal and J. Zuzanek)

and Recreation Service. After the abolition of the Bureau of Outdoor Recreation and the Heritage Conservation and Recreation Service by the middle of 1981, the task of conducting the NRS fell to the National Park Service (NPS) within the US Department of the Interior. The NPS coordinated a consortium of four agencies to sponsor and develop the NRS, including the United States Department of Agriculture Forest Service, the Department of Health and Human Service's Administration on Aging, and the Bureau of Land Management, in addition to the NPS itself.

THE NATIONAL SURVEY ON RECREATION AND THE ENVIRONMENT

By the late 1980s, it was clear that the National Park Service could not afford the financial and organizational demands of a large national survey, so officials informally asked whether the Forest Service would assume the coordinating role for the next NRS. Part of the research branch of the Forest Service, the Outdoor Recreation and Wilderness Assessment Group, agreed to assume this role jointly with the National Oceanic and Atmospheric Administration (NOAA). To raise funds for the survey and to attract a wide range of expertise, the Forest Service and NOAA sought sponsors, including federal and state agencies and private organizations. The final list of sponsoring agencies from the federal government included:

- USDA Forest Service,
- USDI Bureau of Land Management,
- the US Army Corps of Engineers,
- the US Environmental Protection Agency,
- the USDA Economic Research Service (ERS).

The NOAA discontinued its sponsorship shortly before data collection began. In addition to the government sponsors, the Sporting Goods Manufacturers Association, a private recreation industry-sponsored organization, also joined as a sponsor. In-kind resources were offered by the NPS, the University of Georgia and Georgia Southern University, and Indiana University co-sponsored the section on persons with disabilities.

The current name for the 1994 NRS, the *National Survey on Recreation and the Environment* (NSRE), was coined to reflect the growing interest and emphasis on how people of the United States view their natural environment. In comparison with previous NRS models, the scope of the NSRE has been expanded to include more issues related to outdoor recreation participation and natural resources.

The list of sponsors with their variety of information needs, in addition to the central aims of the NRS series, created a special challenge to scientists working on the NSRE. This principal challenge was to provide the wide array of information needed by the sponsors. A relatively small budget, along with time constraints on questionnaire length, made the challenge even greater.

DESIGN AND IMPLEMENTATION

Principal Objectives and Intended Uses

The NSRE describes and explores the participation of people in the United States in outdoor recreation activities, their wildlife and wilderness use and values and attitudes regarding recreation policy issues, and the outdoor recreation participation patterns and needs of people in the United States with challenging and disabling conditions.

NSRE data are generating information for a wide variety of uses. Central among these is the estimation of proportions and numbers of the population participating in outdoor recreation activities. A second purpose is to provide data necessary to produce an estimation of the economic value of outdoor recreation experiences and to develop demand projection models.

The Forest Service will use the data in its *National Assessment of Outdoor Recreation and Wilderness*, which it is required to conduct under the Forest and Rangeland Resources Planning Act. NSRE data will also be used to assist recreation planners and managers at the federal and state level, and will guide policy decisions in land and water management issues. Other uses of the data include the assessment of the role of local, state, federal, and private providers of outdoor recreation, and methods of financing the management of outdoor recreation areas. NSRE data will also be used to generate information about future outdoor recreation markets. University researchers and graduate students will use the data to develop and test theoretically grounded hypotheses, and specialized analyses will provide a range of information relating outdoor recreation participation to a host of theoretically related variables.

Organization of Questions

The NSRE consisted of two separate telephone surveys. The primary survey consisted of a national sample of 12,000 people, aged 16 and over. In interviews lasting on average some 20 minutes, information was gathered on: participation in specified recreation activities measured in number of days and trips; characteristics of recreation trips; barriers and constraints to outdoor recreation; and alternative recreation user fee strategies.

For the secondary survey, a national sample of 5000 people, aged 16 and over, was asked about:

1. participation in outdoor recreation activities;
2. benefits of participation;
3. favourite activities and barriers and constraints to participation;
4. wilderness issues;
5. wildlife issues;
6. awareness of public land agencies; and
7. freshwater-based trips.

All respondents to the secondary survey were asked modules (1) and (2), and were randomly assigned three of the remaining five modules. For each of these randomly assigned modules, the sample size was approximately 2500.

Table 13.1. Activities examined in the US National Survey on Recreation and the Environment.

Running/jogging	Caving
Golf	Bird-watching
Tennis	Wildlife viewing
Baseball	Fish viewing
Basketball	Nature study/photography
Softball	Small game hunting
American football	Big game hunting
Soccer	Migratory bird hunting
Volleyball	Ice skating
Handball/racquetball/squash	Snowboarding
Bicycling	Downhill skiing
Horseback riding	Cross-country skiing
Picnicking	Snowmobiling
Family gathering	Sledding
Yard games/horseshoes, croquet	Off-road vehicle use
Nature museums, zoos	Sightseeing
Visitor centres	Visit beach/waterside
Outdoor concerts/plays	Nature study/water surrounding
Outdoor sports events	Fishing: Freshwater
Prehistoric/archaeological site	Fishing: Saltwater
Visiting a historic area	Fishing: Anadromous
Walking	Fishing: Catch and release
Day hiking	Sailing
Orienteering	Canoeing/kayaking
Backpacking	Rowing
Camping/primitive and developed	Rafting/tubing/other floating
Mountain climbing	Motor boating
Rock climbing	Water-skiing
Jet skiing	Sailboarding/windsurfing
Surfing	Swimming/pool
Swimming/non-pool	Snorkelling

Within these two surveys, participation questions were asked about 62 outdoor recreation activities, as shown in Table 13.1. For 31 of these activities, the number of participation days and the number of recreational trips where the activity was the primary purpose for the trip were collected. Additional information was asked about resource-related issues, such as wilderness and wildlife. For both the primary and secondary surveys, an additional set of questions was used to collect information about disabled user access to recreation areas. These questions were asked only of respondents who indicated that they had a disability. If a respondent indicated that a disabled person lived in the home, the disabled person also was interviewed.

Sampling Framework

The NSRE serves many different purposes and thus had to include a sampling framework to address these purposes. The demand models and estimates of economic value, for example, are assessed from the perspective of the destination of recreational trips, so careful sample design was required to assure adequate numbers of observations covering all destination regions of the country. For the primary survey, a sample stratified by region was employed; within each region, sampling was distributed within states proportionate to the distribution of population among area and local phone codes. Eight regions were identified, as shown in Figure 13.1.

To ensure adequate numbers of observations (minimum of 900 per region and 400 for Alaska) in the Rocky Mountains, the Great Plains and Alaska, a disproportionate sampling rate with respect to population proportion was used, as shown in Table 13.2. For the secondary survey, a simple random sample of the nation's population was employed. This sample was distributed among the states proportionate to population distribution. In addition, the data were post-weighted for analysis to compensate for disproportionate sampling rates with respect to social strata and geographic regions.

Participation Questions

Because the NSRE is designed to serve many different purposes, the level of detail needed to describe participation in activities varies. For each activity, a categorical yes/no response was collected to represent whether the respondent did or did not participate in an activity at least once within the past 12 months. For selected activities, frequency (number of days) of participation was also collected. For activities for which demand models were to be generated, additional data were collected to identify how many trips away from home were undertaken when the activity was the primary purpose for

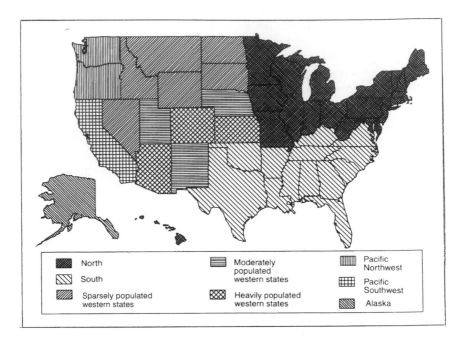

Fig. 13.1. US National Survey on Recreation and the Environment: sampling regions.

the trip. The activities for which these 'primary purpose trips' were taken were stored by the Computer Assisted Telephone Interview system and one activity among those identified was randomly selected by the computer for additional trip characteristic questions.

Table 13.2. Sampling, estimated regional sample sizes, US National Survey on Recreation and the Environment, 1994.

Region	Per cent of population	Allocated sample size	Total sample	Sample %
North	46.9	1,000	2,897	24.1
South	30.3	1,000	2,229	18.6
Sparsely populated western states	2.0	1,800	1,800	15.0
Moderately populated western states	1.8	900	900	7.5
Heavily populated western states	3.9	1,300	1,300	10.8
Pacific Northwest	3.3	1,000	1,000	8.3
Pacific Southwest	11.7	1,000	1,474	12.3
Alaska	0.2	400	400	3.3
Total*	100.0	8,400	12,000	100.0

* Total does not add up to 100 due to rounding.

Issue Specific Sections

As described earlier, the NSRE was a multi-sectional survey focusing on a variety of recreational issues. Short descriptions of a number of such issues covered in the survey are provided below.

Persons with disabilities
A very significant issue in the United States is whether persons with challenging conditions are inappropriately restricted from access to outdoor recreation areas. In addition to general access concerns, the NSRE included a section on wilderness access for the physically challenged. Access questions were included to address both legislatively mandated and policy-driven programmes which seek to improve access for all United States citizens. Because this sub-sample of respondents was asked the standard NSRE questions as well, the survey provides the most comprehensive national outdoor recreation profile of persons with disabilities in the United States yet assembled.

Wilderness and wildlife resource policy issues
Despite numerous studies of wilderness users, the general American public has been little studied with regard to its values and awareness of wilderness resources. Similarly, non-economic wildlife values have been inadequately studied. Coupled with other sections of the questionnaire, for example, participation profiles and demographics, specifically tailored questions about these natural resource issues were included to provide much needed information to guide policy evaluation and management.

Trip profiles and valuation objectives
Resource economics literature describes a method generally referred to as travel cost modelling (Clawson and Knetsch, 1962). This methodology focuses on recreational trips taken to different types and qualities of destination sites. Because persons taking such trips are willing to incur costs in taking trips, a relationship between cost and number of trips taken can be observed. From this relationship, a formal trip demand function can be estimated and *consumer surplus* calculated. Consumer surplus is the amount consumers are willing to pay for a recreational trip over and above what they actually do pay. This estimate has been shown to be a theoretically valid measure of the value of recreational opportunities (Walsh, 1986; Bergstrom, 1990). To gather data for this type of analysis NSRE respondents were asked whether they took one or more trips away from home to participate in outdoor recreational activities. For those reporting in the affirmative, a bank of questions was asked about the last such recreation trip.

Favourite activities

Because individuals vary in what they enjoy and commit themselves to in outdoor recreation, a section of the survey asked about favourite activities. Included was a measure of commitment and the preferred 'setting' or environment for the identified favourite activity. Learning theory posits that *self-directed* learning may be different from *other*-directed learning. For example, variation in participation in formal, informal, and non-formal learning environments may be related to psychological and social states of the participant (Candy, 1987; Pratt, 1988). Questions in the NSRE were designed to address these issues.

Barriers and constraints

Reasons for non-participation in outdoor recreation are of particular interest to outdoor recreation managers. The NSRE replicated and expanded the list of barriers and constraints from the 1982–83 NRS and allowed open-ended responses to capture any new or previously unidentified barriers and constraints. This section was asked in one of two situations: (i) for respondents who reported that they did not participate in any outdoor recreation, and (ii) for respondents who reported that they did not participate in their favourite activity as often as they would have liked.

Reports Anticipated

A number of NSRE reports are planned. A general summary report will be produced for widespread distribution, generally following the format of the 1982–83 NRS report (US Department of the Interior, 1986). This summary report will display the data descriptively and cover all major areas of the NSRE, but will emphasize participation. Additional reports will include participation differences among demographic strata, participation by older people and people with disabilities, barriers and constraints to participation, and favourite activities. Other reports will focus on subsets of the data to analyse specific issues, for example, wildland recreation and resource management, emerging markets, and outdoor recreation travel. Another report will document and evaluate the NSRE methodology.

METHODOLOGICAL ISSUES

Non-response Bias and Control

There are many potential sources of bias in a large survey of human subjects. The principal ones of concern for NSRE include recall and 'digit preference' among the response biases, and refusal, avidity, and incomplete telephone listings among the non-response biases.

Recall bias is simply an inability of a respondent to recall accurately, or to recall at all, whether they participated in recreational activities, or to recall the number of or places where these activities were undertaken. There is no conclusive evidence regarding optimum recall period (1 week, 1 month, 6 months, etc.) or methods for correcting recall bias. A calibration study is anticipated for the NSRE to solicit participation from a panel of respondents at 1-month and 12-month intervals in order to compute correction weights for the principal NSRE sample.

Digit preference bias is related to recall bias, but more specifically is a participation rounding bias. For example, for activities of frequent participation, such as walking or running/jogging, respondents often round to the nearest five or ten, such as 25, 30 or 40, rather than accurately reporting actual number of occasions, such as 28 times during the last 12 months.

Principal sources of non-response bias include avidity and incomplete telephone listings. Avidity bias is the tendency of persons who do not participate in outdoor leisure activities, or who participate only infrequently, to refuse to participate in the survey. Left unaccounted for, avidity bias can result in seriously inflated estimates of population participation rates and biased estimates of participation differences by social group. Incomplete telephone listings, like any other incomplete sampling frame, can occur for many reasons. More frequently encountered reasons include institutionalization, persons not having a phone, and persons having access only to pay phones or other non-individualistic arrangements. For the NSRE, an attempt to estimate avidity and listing bias was made by asking two key questions of persons who refused the survey. Those questions are age and whether or not the respondent had participated in outdoor recreation in the last 12 months. Additionally, the gender of the respondent was recorded when recognizable. The estimated proportions of non-respondents, relative to respondents, was combined with weights derived from the 1990 Census of the US population to weight each observation to correct for over-representation or under-representation by that respondent's social group in the sample.

As with any survey, regardless of scope or complexity, bias is a reality to be recognized and dealt with early on, to the extent affordable, through design of the sample and the questionnaire content.

Definitions and Understanding of Activities

The NSRE included a more comprehensive listing of outdoor recreation activities than any of the previous national surveys. The activities list for the NSRE included 62 explicitly named activities. Some of these activities are, and always have been, relatively vague, for example sightseeing and walking for pleasure. Others are much more specific and have relatively precise technical definitions, for example snorkelling and rock climbing. Respondents are left to determine, according to their own definition of these activities, whether or not they have participated. For the NSRE, several new activities were listed, largely driven by newly available or significantly improved technologies, such as mountain biking, jet-skiing, rock climbing and orienteering. To the extent that respondents understand the activities they are being asked about, valid responses are recorded. However, little exists in the literature to guide control for this potential source of error in collecting data on participation.

Valuation Model Specification and Geo-positioning of Destinations

In relation to the gathering of data to facilitate the travel cost modelling methodology referred to above, significant methodological issues include the problem of measuring trip cost and quality of the recreational destination, and assessment of the availability of substitute destinations, congestion effects, and socioeconomic differences in tastes. A number of these demand factors depend on explicit identification of the recreational destination. Accurate identification of these destinations is, however, hampered by respondent misunderstanding of site name, lack of knowledge of site location, and lack of knowledge of resource owner or site manager.

PATTERNS OF PARTICIPATION

Overall Participation

At the time of writing, only part of the planned data had been collected. The following findings are based on a sample of 7504 respondents. The final column of Table 13.3 shows the proportion of the adult US population participating in any outdoor recreation activity and in each of the top 38 activities included in the survey, in the last 12 months. Over 80% of respondents had taken part in at least one of the activities at least once in the previous 12 months. Activities with highest levels of participation include walking for pleasure, attending family gatherings outdoors, picnicking, visiting a beach or waterside, sightseeing, attending outdoor sporting events, and swimming in pools. Least participated in are orienteering, handball and squash, backpacking, caving, rock climbing, and snorkelling.

Table 13.3. Activity participation by gender, USA, 1994.

	% participating in previous year		
	Males	Females	Total
Outdoor recreation participation	–	–	94.5
Running/jogging	31.6	21.1	26.2
Walking	65.1	68.5	66.7
Golf	22.4	7.8	14.8
Tennis	11.7	9.6	10.6
Outdoor team sports	31.2	18.6	24.7
Handball/racquetball/squash	6.2	5.1	5.6
Bicycle	31.0	26.5	28.6
Horseback riding	7.0	7.3	7.1
Picnic	47.1	51.0	49.1
Family gatherings	63.8	60.2	61.8
Yard games	41.0	32.9	36.9
Nature museum	46.3	46.7	46.4
Visitor centres	36.0	33.4	34.6
Outdoor concerts/plays	21.7	19.9	20.7
Outdoor sports events	53.6	42.0	47.5
Prehistoric structure/archaeological sites	19.5	15.6	17.4
Historic areas/sites/buildings/memorials	46.1	42.5	44.1
Hiking	27.1	20.9	23.8
Backpacking	10.2	5.1	7.6
Developed camping	22.9	18.7	20.7
Primitive camping	19.0	9.4	14.0
Mountain climbing	5.8	3.3	4.5
Rock climbing	5.1	2.5	3.7
Orienteering	3.6	1.3	2.4
Bird-watching	24.7	29.2	27.0
Wildlife viewing	31.4	31.2	31.2
Hunting	16.8	2.7	9.4
Caving	5.7	3.8	4.7
Off-road driving	17.9	10.3	13.9
Sightseeing	57.8	55.7	56.7
Snow and ice activities	22.0	17.1	19.4
Visit beach/waterside	64.9	59.6	62.1
Fishing	37.8	21.1	29.1
Study nature/near water	28.4	27.0	27.6
Boating	34.3	26.2	30.0
Swimming/in pool	45.6	43.1	44.2
Swimming/other than pool	42.8	35.6	39.0
Snorkelling	9.1	5.5	7.2

Participation Profiles by Social Characteristic

Among the most popular activities, participation rates are relatively uniform across gender and age groups, as shown in Tables 13.3 and 13.4. Activities exhibiting the greatest participation difference by gender include running/ jogging, golf, outdoor team sports, backpacking, orienteering, and hunting. Across all of these activities, males reported participation more frequently than females.

Activities exhibiting the greatest age differences include running/jogging, tennis, outdoor team sports, bicycling, horseback riding, and snow and ice activities. For all of these activities, participation drops rapidly with rising age. Activities for which participation rates remain relatively steady or increase across age groups include visiting nature museums, touring visitor centres, visiting prehistoric and historic sites, bird watching, wildlife viewing, studying nature near water, hunting, sightseeing, and fishing.

A sufficient sample was collected to examine recreation patterns of Caucasians, African Americans and a category designated as 'other' Americans (including Hispanics, Native Americans and Asian/Pacific Islanders), as shown in Table 13.5. Recreation participation appears to vary in relation to race, with Caucasians generally reporting higher levels of participation. African Americans are the least represented in virtually all recreation activities, with the notable exception of outdoor team sports, running and jogging and handball/racquetball/squash. In particular, African Americans did not report participation in the less developed outdoor recreation activities, such as primitive camping, hiking and hunting.

Regional variation in participation, as shown in Table 13.6, is marked for some activities, reflecting such factors as climate and tradition, but there is no one region which is more active overall.

Income has a very strong effect on participation across almost all activities, as shown in Table 13.7. Those activities for which greater income has the strongest effect in increasing participation include golf, tennis, visiting prehistoric and archaeological sites, hiking, rock climbing, studying nature near water, and snorkelling. Those activities for which participation rates remain relatively steady across income groups include hunting, off-road motor vehicle driving, fishing and outdoor swimming in natural waters. Overall, however, participation was strongly and positively correlated with rising income, rising from 62% among those earning less than $10,000 to 94% among those earning over $100,000.

Table 13.4. Activity participation by age, USA, 1994.

Age (in years)	% of age-group participating in previous year						
	16–24	25–29	30–39	40–49	50–59	60+	Total
Outdoor recreation participation	–	–	–	–	–	–	94.5
Running/jogging	50.4	33.3	28.4	23.3	17.2	8.0	26.2
Walking	68.2	72.5	74.7	72.0	65.5	51.8	66.7
Golf	15.3	19.0	17.7	15.4	12.0	10.3	14.8
Tennis	21.4	12.2	11.7	9.7	6.1	3.1	10.6
Outdoor team sports	47.4	37.2	30.2	20.2	12.9	4.0	24.7
Handball/racquetball/squash	12.5	7.3	5.8	3.9	3.8	1.5	5.6
Bicycle	38.0	36.3	37.4	30.6	21.9	10.7	28.6
Horseback riding	12.4	10.2	8.8	7.2	4.3	1.2	7.1
Picnic	45.1	54.1	59.8	55.4	47.7	35.1	49.1
Family gatherings	68.1	70.2	69.2	65.4	59.5	44.1	61.8
Yard games	43.6	46.4	47.1	39.4	31.3	16.9	36.9
Nature museum	44.2	54.9	56.6	52.4	44.2	30.5	46.4
Visitor centres	29.6	37.1	41.1	41.8	37.8	24.0	34.6
Outdoor concerts/plays	26.8	26.3	22.6	20.6	17.8	13.0	20.7
Outdoor sports events	59.5	51.1	54.0	52.5	43.6	28.2	47.5
Prehistoric structures/archaeological sites	18.5	20.0	19.3	19.8	17.4	11.8	17.4
Historic areas/sites/buildings/memorials	44.6	48.7	49.8	50.4	45.4	30.8	44.1
Hiking	31.5	30.2	29.5	27.0	18.1	9.6	23.8
Backpacking	14.3	11.9	8.3	7.0	4.5	1.5	7.6
Developed camping	27.6	25.5	26.0	22.6	15.6	9.0	20.7
Primitive camping	22.7	20.0	16.6	13.8	9.2	4.3	14.0
Mountain climbing	8.2	6.3	5.3	3.7	2.3	1.7	4.5
Rock climbing	8.3	5.5	3.9	2.9	1.8	0.7	3.7

Orienteering	4.1	3.2	2.7	2.4	1.5	0.9	2.4
Bird-watching	16.4	21.2	28.4	33.8	32.8	28.9	27.0
Wildlife viewing	28.8	32.3	36.5	37.7	32.8	21.9	31.2
Hunting	13.4	12.2	11.1	9.1	9.0	3.8	9.4
Caving	7.9	7.0	5.3	4.3	2.9	1.6	4.7
Off-road driving	23.4	19.1	15.3	12.3	10.4	5.6	13.9
Sightseeing	52.0	60.4	63.4	64.4	59.6	44.5	56.7
Snow and ice activities	28.4	26.1	26.0	20.4	13.2	4.9	19.4
Visit beach/waterside	70.2	71.6	71.3	66.9	57.1	40.9	62.1
Fishing	35.4	34.3	33.9	31.5	24.9	17.1	29.1
Study nature/near water	25.7	30.7	32.7	34.6	29.1	16.6	27.6
Boating	37.2	37.8	34.0	30.9	26.8	17.7	30.0
Swimming/in pool	60.6	54.7	53.0	44.8	34.8	21.8	44.2
Swimming/other than pool	51.3	49.5	48.2	42.4	30.5	16.7	39.0
Snorkelling	10.4	10.7	9.4	8.1	4.7	1.4	7.2

Table 13.5. Activity participation by race, USA, 1994.

| | \% participating in last year | | | |
	Caucasian	African-American	Other	Total
Outdoor recreation participation	–	–	–	94.5
Running/jogging	24.9	31.9	31.7	26.2
Walking	69.0	55.8	58.6	66.7
Golf	16.9	4.1	7.1	14.8
Tennis	10.8	7.8	12.9	10.6
Outdoor team sports	23.6	31.4	25.1	24.7
Handball/raquetball/squash	4.9	8.5	10.4	5.6
Bicycle	29.9	21.2	26.2	28.6
Horseback riding	7.6	3.7	6.5	7.1
Picnic	57.7	42.4	70.3	49.1
Family gatherings	63.2	56.0	56.2	61.8
Yard games	39.2	26.3	24.9	36.9
Nature museum	48.7	33.2	42.0	46.4
Visitor centres	36.8	22.8	28.6	34.6
Outdoor concerts/plays	21.2	17.3	20.7	20.7
Outdoor sports events	49.3	38.5	40.9	47.5
Prehistoric structure/archaeological sites	18.0	13.3	18.1	17.4
Historic areas/sites/buildings/memorials	46.6	31.1	36.8	44.1
Hiking	26.1	23.4	8.1	23.8
Backpacking	8.2	7.5	3.0	7.6
Developed camping	22.9	6.3	18.2	20.7
Primitive camping	15.5	4.2	12.3	14.0
Mountain climbing	4.8	1.9	4.9	4.5
Rock climbing	4.2	0.4	3.3	3.7
Orienteering	2.4	1.7	3.3	2.4
Bird-watching	28.4	20.5	21.4	27.0
Wildlife viewing	33.6	18.1	25.1	31.2
Hunting	10.6	3.4	5.7	9.4
Caving	5.1	1.7	4.8	4.7
Off-road driving	14.7	8.3	14.0	13.9
Sightseeing	59.5	41.9	47.6	56.7
Snow and ice activities	21.4	7.9	14.6	19.4
Visit beach/waterside	64.8	46.0	56.9	62.1
Fishing	30.9	19.6	23.7	29.1
Study nature/near water	29.5	15.5	25.8	27.6
Boating	33.5	10.5	20.3	30.0
Swimming/in pool	47.0	29.1	36.1	44.2
Swimming/other than pool	42.6	16.8	32.2	39.0
Snorkelling	7.8	3.0	7.5	7.2

H.K. Cordell et al.

Table 13.6. Activity participation by region, USA, 1994.*

	% participating in previous year			
	Northeast	Midwest	South	West
Outdoor recreation participation	–	–	–	–
Running/jogging	25.4	23.4	27.2	28.5
Walking	68.2	68.3	64.3	67.8
Golf	13.8	18.6	12.5	15.0
Tennis	11.5	9.6	10.7	10.5
Outdoor team sports	24.6	25.5	24.9	23.1
Handball/racquetball/squash	6.4	4.7	5.4	6.4
Bicycle	28.2	32.4	24.6	31.5
Horseback riding	5.4	6.8	7.4	8.7
Picnic	49.5	52.3	44.8	52.2
Family gatherings	60.6	65.7	59.7	62.3
Yard games	40.4	40.8	34.7	31.9
Nature museum	44.3	50.5	42.9	50.0
Visitor centres	36.1	32.8	33.4	37.0
Outdoor concerts/plays	19.2	19.7	18.3	27.7
Outdoor sports events	47.6	48.5	47.8	46.0
Prehistoric structure/archaeological sites	15.2	16.9	16.3	22.3
Historic areas/sites/buildings/memorials	44.9	43.9	43.7	44.9
Hiking	21.8	22.6	18.6	36.2
Backpacking	7.6	5.4	5.9	12.8
Developed camping	17.6	21.8	17.2	28.4
Primitive camping	10.1	13.7	12.5	20.7
Mountain climbing	4.4	2.6	3.6	8.4
Rock climbing	3.1	3.2	2.9	6.1
Orienteering	2.8	2.1	2.1	2.9
Bird-watching	27.9	29.2	26.2	24.9
Wildlife viewing	30.5	34.0	28.9	32.6
Hunting	6.7	11.5	10.8	7.5
Caving	2.6	5.4	4.5	6.1
Off-road driving	11.1	12.6	14.7	17.0
Sightseeing	56.5	57.6	54.4	59.7
Snow and ice activities	26.5	24.1	10.8	21.5
Visit beach/waterside	64.4	61.4	60.4	63.8
Fishing	23.9	31.8	32.2	25.8
Study nature/near water	28.3	26.2	26.6	30.5
Boating	29.1	32.9	30.0	27.8
Swimming/in pool	47.6	42.3	44.9	42.4
Swimming/other than pool	44.9	39.1	37.4	36.1
Snorkelling	7.2	5.2	7.7	9.0

* **Northeast**: Maine, New Hampshire, Vermont, Massachusetts, Rhode Island, Connecticut, New York, New Jersey, and Pennsylvania. **Midwest**: Ohio, Indiana, Illinois, Michigan, Wisconsin, Minnesota, Iowa, Missouri, North Dakota, South Dakota, Nebraska, and Kansas. **South**: Delaware, Maryland, District of Columbia, Virginia, West Virginia, North Carolina, South Carolina, Georgia, Florida, Kentucky, Tennessee, Alabama, Mississippi, Arkansas, Louisiana, Oklahoma, and Texas. **West**: Montana, Idaho, Wyoming, Colorado, New Mexico, Arizona, Utah, Nevada, Washington, Oregon, California, Alaska, and Hawaii.

Table 13.7. Activity participation by income, USA, 1994.

Annual personal gross income, US$'000s	% participating in previous year				
	< $25	$25–50	$50–75	$75–100	$100+
Outdoor recreation participation	–	–	–		–
Running/jogging	18.6	26.4	29.6	30.9	38.1
Walking	49.9	71.9	74.9	77.1	74.8
Golf	5.7	14.5	22.6	24.6	30.0
Tennis	6.3	9.9	13.8	15.3	19.2
Outdoor team sports	16.3	27.6	27.6	27.1	24.3
Handball/racquetball/squash	4.9	5.0	5.6	6.9	6.4
Bicycle	18.0	31.1	35.9	37.0	39.1
Horseback riding	3.7	7.4	9.6	9.9	11.2
Picnic	35.8	55.5	56.6	56.3	51.8
Family gatherings	43.7	68.1	70.7	70.1	71.9
Yard games	22.5	41.8	45.4	44.1	39.2
Nature museum	28.8	51.6	58.7	61.5	59.4
Visitor centres	21.4	37.9	45.2	45.0	46.1
Outdoor concerts/plays	14.3	21.2	26.2	26.2	30.4
Outdoor sports events	27.4	49.4	60.0	62.6	64.2
Prehistoric structure/archaeological sites	10.8	18.9	21.6	21.0	24.6
Historic areas/sites/buildings/memorials	26.3	48.2	56.3	56.5	60.3
Hiking	16.2	26.2	30.7	30.5	32.8
Backpacking	5.3	7.8	9.9	8.7	10.1
Developed camping	14.8	24.4	25.7	22.8	19.6
Primitive camping	10.1	16.3	16.6	14.9	14.1
Mountain climbing	2.8	4.6	5.9	6.1	6.7
Rock climbing	2.4	3.6	5.7	2.8	5.1
Orienteering	1.2	2.6	3.2	2.7	3.1
Bird-watching	20.8	28.6	29.9	32.3	29.3
Wildlife viewing	20.4	35.1	37.2	40.4	35.5
Hunting	4.4	11.6	10.8	10.5	8.0
Caving	3.3	5.1	5.5	5.2	5.1
Off-road driving	9.7	16.3	16.8	13.6	16.8
Sightseeing	37.0	62.7	68.6	70.3	70.4
Snow and ice activities	10.4	21.0	25.9	28.8	31.4
Visit beach/waterside	39.3	66.7	74.5	79.2	78.6
Fishing	20.3	32.2	33.9	33.7	32.1
Study nature/near water	15.3	30.0	35.8	40.2	35.5
Boating	15.6	32.2	39.5	41.9	43.1
Swimming/in pool	25.2	48.2	55.6	60.1	59.6
Swimming/other than pool	23.7	42.3	48.6	52.7	54.7
Snorkelling	2.5	6.1	11.0	14.0	17.6

SOCIAL AND POLITICAL IMPLICATIONS

The United States is rapidly becoming a truly multi-cultural society, in terms of race and ethnicity, characteristics of residence, family structure, education, religion, and values. From the perspective of outdoor recreation, it is important to view participation in light of these social and cultural influences. The NSRE seems to represent a diversity of cultures to the extent that the US population also represents a diversity.

The NSRE data indicate that single parents participate far less in outdoor recreation than parents from dual parent households. While not a surprising finding, it indicates that socioeconomic factors related to single parenthood, such as income and available leisure time, may have a negative impact on a single parent's recreation participation. Other issues of concern in the United States include sociocultural bias in the presentation and management of areas such as prehistoric, historic, and cultural sites; issues of physical and programmatic access; and inequality of access based on age, income, place of residence, and other factors. Non-Caucasian visitors to wildland recreation areas are still significantly under-represented, raising questions about awareness, access, culture, values and, potentially, discrimination. For example, African Americans tend not to participate in activities such as camping, rockclimbing, or caving, but participate more in developed activities such as walking, American football, and outdoor team sports.

Continued growth of the environmental movement has created new awareness of formerly unrecognized sociocultural influences on outdoor recreation, including a range of values concerning the use and management of the nation's natural resources. Research questions will focus on the potential influence that these sociocultural factors and values may have on outdoor recreation participation behaviours and choices. The NSRE can illuminate our understanding of the potential sociocultural influences on recreation participation.

On 11 September 1993, President Clinton issued an Executive Order to federal agencies requiring each to develop standards for customer service. This mandate requires agencies of the federal executive branch to survey customers, establish baselines of customers, and monitor progress in being more responsive to customer demands and providing quality service. Recreation users of federal and other public lands are among the most visible, and certainly most numerous, among customers of the land management agencies. Understanding the participation preferences, barriers and constraints to participation, favourite activities and values associated with recreational experiences of the United States public is central to customer service responsiveness. The NSRE provides the United States with a crucial baseline of information and analytical capability for tracking trends and evaluating current conditions in outdoor recreation service. Several current, more specific policy issues may also be addressed using the NSRE database.

Serious concern has emerged in the United States over the last few years that government has become so complicated and inward looking that it has, for all intentions and purposes, ceased to function effectively. Part of the mandate likely to surface as proposals for reinvention of the federal bureaucracy is that decisions must be science-based. The NSRE provides a theoretically based data set for understanding the demand for outdoor recreation and for estimating its market value. Understanding market demand and value is useful for comparison with other potentially competing uses of natural areas managed by the federal sector. While demand and estimated value are not the only inputs for decision making, these measures of the relative importance of outdoor recreation enable a science-based comparison and evaluation relative to other potential uses of natural resources.

Since the turn of the century and before, management of natural resources, particularly forests, has been based predominantly on commodity values. Application of biological and physical sciences was primarily aimed at improving technologies for management of natural resources to increase the quality and quantity of commodity output. In more recent years, non-commodity human values of these same natural resources (some of which are no longer truly 'natural') have played increasingly prominent roles. Personal, community, and ecosystem values are now considered, along with commodity values, in deciding the future of natural lands and waters. Consequently, social sciences and projects such as the NSRE are playing a much more significant role in providing the background for policies and plans. In the general shift of concern toward health of ecosystems, as opposed to the commodity outputs from these ecosystems, humans are considered as legitimate components. As such, social scientists are now being integrated into research teams looking at the condition, productivity, and desired future condition of ecosystems in the United States.

Setting policy and the pursuit of political agendas in contemporary United States culture requires careful attention to the variety of different values associated with natural resources and outdoor spaces. The NSRE was designed to sample and understand some of these differences. Only recently have we begun to understand that the reasons for recreational choices and styles of participation are much more complex than just differences of socioeconomic status and education. Perhaps even more important is the cultural background of the people making those recreational choices, or choosing not to participate. The United States, since the coming of the Europeans, has always been composed of a variety of people from a variety of cultures. This diversity is increasing even more as the pressures of world population cause significant migrations. The NSRE will shed light on that portion of this diversity pertaining to outdoor recreation.

CONCLUSIONS

The National Survey on Recreation and the Environment represents the most recent data point in a continuing series of national surveys in the United States. As such, it offers insights into one dimension of change which the country is undergoing as we near the year 2000. As society changes more rapidly with each passing decade, there is a need to continue these surveys into the future and a need to stay in touch with, and be sensitive to, institutional changes within which survey results must be presented and interpreted.

Changing ethnic characteristics, values and, most importantly, choices and behaviours, will demand a different response in the future than was the case in the past. Tremendous pressures will be placed on natural resources in coming years as the population of the United States increases by nearly 50% by the year 2050. Planning for these coming years will require solid information on the many uses and demands the public places, and will be placing, on these precious resources of land and water. Recreation demands and uses will present some of the most highly pressured demands on this resource base. Gone are the days when guessing will suffice; the United States public is very much aware of resource opportunities and threats to their access to them.

Ideally the National Survey on Recreation and the Environment, as an on-going survey series, should retain comparability through a core of questions on participation, barriers to participation and related issues. This would create a baseline against which to measure the magnitude and composition of change. In addition to this baseline data, special areas of enquiry will need to be incorporated as issues emerge and situations change. Tied back to this baseline of participation, barriers, preferences and demographics, these special areas of enquiry would enable enhanced understanding of the sources of change and concern, and perhaps some interpretation of their causes.

From the beginnings of resource management in the United States, biological sciences were practically the sole focus. Forests were viewed as sources of raw materials and parks as scenic and biological preserves. In the last few years, however, a revolution in thinking and values has taken place wherein social sciences have begun to be viewed as a legitimate partner in resource evaluation. For example, most federal resource management agencies in the United States are now adopting ecosystem management as the principal doctrine guiding management priorities. But ecosystems in this changing institutional environment are not viewed solely as biological and physical systems; the human dimension of ecosystems is also included. In this context, the social sciences play a prominent role in assessments of ecosystem health, status and sustainability.

With this more 'legitimate' and prominent posture, social scientists must have valid and up-to-date databases from which to work. Surveys such as the

NSRE are among the most important sources of human dimensional data. The outdoor recreation activity which is the focus of the NSRE, for the most part, occurs within the ecosystems of interest. Behaviours and values of outdoor recreation participants are crucial elements of the human dimension and represent the most direct interface that resource managers have with the public on whose behalf ecosystem management is undertaken.

In past years, the desired goal was to update the National Recreation Survey every 5 years. Because of limited resources and some loss of interest among the original sponsors of the national surveys, it seems that updates every 8–10 years are a more realistic possibility. A limitation imposed on the current NSRE was that there was no central source of funding identified solely for the core of the survey, that is participation, items related directly to participation, and demographics. Multiple sponsorship led to an extremely complex survey designed to meet the many needs of multiple sponsors. Ideally, future NSRE efforts should be undertaken only after a dominant funding source is made available, which is directed at the core of the national survey. Other issues or areas of enquiry should be undertaken only if they do not hamper accomplishment of the core.

NOTE

1. Additional material on United States surveys can be found in some of the comparative data in Chapter 3, on Canada, pp. 63–65.

REFERENCES

Bergstrom, J.C. (1990) Concepts and measures of economic value of environmental quality: a review. *Journal of Environmental Management* 31, 215–228.

Candy, P.C. (1987) Reframing research into 'self direction' in adult education: A constructivist perspective. Doctoral dissertation, Department of Adult and Higher Education, University of British Columbia, Vancouver.

Clawson, M. and Knetsch, J.L. (1962) *The Economics of Outdoor Recreation*. Johns Hopkins Press, Baltimore.

Outdoor Recreation Resources Review Commission (1962) *Outdoor Recreation for America*. ORRRC, Washington, DC.

Pratt, D.D. (1988) Andragogy as a relational construct. *Adult Education Quarterly* 38(3) 160–181.

United States Department of the Interior, National Park Service (1986) *1982–1983 Nationwide Recreation Survey*. US Government Printing Office, Washington, DC.

Walsh, R.G. (1986) *Recreation Economic Decisions: Comparing Benefits and Costs*. Venture, State College, PA.

14 Cross-national Leisure Participation Research: A Future

GRANT CUSHMAN, A.J. VEAL AND JIRI ZUZANEK

INTRODUCTION

The contributors to this book were asked to address: first, the experience of national leisure surveys in their respective countries; second, overall patterns of participation including, where possible, trends over time; third indicators of inequalities in participation in relation to such factors as gender, age and socioeconomic status; and fourth, particular problems or issues which the conduct of national surveys had thrown up. In this final chapter these four areas are reviewed in turn, in relation to the prospect of future surveys and their cross-national comparison.

THE EXPERIENCE

The 12 contributions to this book reveal a wide range of experience in the conduct of national leisure participation surveys. Table 14.1 summarizes some of the characteristics of the surveys reported on. With the exception of France, all the surveys were conducted in the 1990s. Sample sizes ranged from as few as 1200 to as many as 19,500, with an average of about 7000. Smaller sample sizes have implications in terms of the reliability of data. It is a reflection of the particular research tradition from which these surveys arise, that none of the chapter contributors provided statistical tests of significance with their data. Since most of the analysis relates to descriptive presentation of

Table 14.1. National survey characteristics.

Country	Date of latest survey	Sample size	Age-range of sample	Reference period
Australia	1991	2,100	14+	One week
Canada	1992	12,000	12+	One year
France	1988	10,870	14+	Varies
Germany	1991	3,069	14+	Often
Great Britain	1990	19,500	16+	4 weeks
Hong Kong	1993	2,917	6+	One month
Israel	1990	1,197	20+	One day
Japan	1992	3,408	15+	One year
New Zealand	1991	4,373	15+	4 weeks
Poland	1990	11,700	15+	1 year
Spain	1992	8,500	18+	Not specified
United States	1994	12,000	16+	One year

results from a single survey this is perhaps not surprising. However, smaller sample sizes present problems when more detailed analyses and comparisons between surveys are attempted.

The 'age threshold' of the samples covered by the surveys varies substantially from country to country, with minimum ages ranging from 6 years to 20 years. In some surveys, therefore, the bulk of teenagers are excluded, and only one includes children under 12 years old. This is highly significant in certain areas of leisure such as active sport, popular music and computer and electronic games, where young people are generally very active. If comparability is to be achieved in future then a common threshold age will need to be adopted – or at least it will be necessary to make sub-sets of the data available covering common age-ranges.

PATTERNS OF PARTICIPATION

While the list of activities covered by the surveys varies enormously, all 12 of the countries included can claim to have comprehensive leisure surveys, including home-based and away-from-home leisure, and sporting, arts, cultural, social and outdoor activities. In some cases the results of time-budget surveys are reported, especially in those countries where these have been an important focus of leisure research, however, it should be noted that the picture is not complete since some countries, for example Australia and Great Britain, have time-budget data available, but they have not been included here.

Table 14.2 shows that a total of 195 activities are listed in the surveys reported in the preceding chapters. No one activity is common to all surveys reported, although television watching and going to the movies come close. In many cases comparability is confounded by the tendency of survey designers to group activities together. Examples include opera and concert going, reading of various types, indoor and outdoor sporting activities and various water-based activities. As Zuzanek points out in the chapter on Canada, groupings of activities may lead to inaccuracies in responses due to the 'inflationary' reporting of participation, where respondents feel that they *must* have engaged in one or more of a group of activities (e.g. 'performing arts'), even if they cannot recall specific instances. One lesson for the future is for the designers of surveys *not* to group activities together, but to retain as much detail as possible, to facilitate comparison across surveys.

The major factor preventing comparison between surveys is the 'reference period' chosen – that is the time period to which reported participation relates. This ranges from as little as a couple of days (Israel) to a year. Clearly the range of activities which an individual engages in during the course of a year is much greater than the range they participate in during a single week or month. The longer the reference period used, therefore, the higher the apparent participation rates. The United States has the most extensive database and it is almost entirely based on a 1-year reference period. The 1-year period has the advantage of covering all seasons, but it can be argued that the scope for error in recalling activities over such a long time period is great. In recent years there has been some attempt to collect data relating to both a shorter and a longer time period in the same survey, and to compare the results (see Table 6.4 in relation to Great Britain and see Darcy (1993) in relation to Australia), but it is too early to draw any general conclusions from these experiments. Before any comparisons can be made across countries, this issue will have to be resolved.

Overall, therefore, it can be said that, because of survey design differences, at this stage, it is *not* possible to compare survey results cross-nationally.

In some countries a series of surveys has been established over a number of years, which enables trends to be established. These include France, Great Britain, Israel, Japan, Poland and the USA. However, in some of these cases there remain problems of comparability between surveys and in a number of cases only two surveys are available, which is not a strong basis for the establishment of trends. The overall picture, in terms of trends, is mixed. Certainly up to the 1980s, leisure time appeared to be increasing, at least for some groups in the community, but there is evidence to suggest that it has declined in the 1990s. There is no clear trend in leisure participation: some activities increase in popularity while others go into decline. As discussed in Chapter 1, most of the surveys are sponsored by government agencies in order to monitor the effects of government policies: in particular governments would hope to find increasing levels of participation in those types of activity,

Table 14.2. Activities covered in surveys.*

Arts/Cultural Activities	AUS	CAN	FRA	GER	GBR	HK	ISR	JAP	NZ	POL	SPA	USA†
Art films/ciné-clubs		●										
Art gallery/art museum		●						●				
Arts crafts	●	●						●				
Arts (paint, sculpt)		●	●		●			●				
Ballet			●		●							
Circus		●	●		●	●		●		●●		●
Classical music concert		●●	●●	●●	●●	●●		●●		●	●	
Flower arranging												
Historic site visit		●	●									
Jazz concert	●	●										
Library visit	●	●	●	●	●	●		●		●	●	●
Movies	●	●	●	●	●	●●				●		
Museum/historic site visit						●						
Museums, galleries	●					●				●		
Music recital/opera	●											
Opera or ballet		●	●		●			●				
Opera	●●	●●	●●	●●	●●	●		●●		●	●	
Other live performances	●	●										
Outdoor concert/play		●		●		●		●		●	●	●
Painting, sculpturing, pottery										●		
Performing arts	●	●●	●	●	●●	●		●●●		●		
Photography						●		●				
Playing musical instrument	●	●		●				●		●		
Playing music		●●	●					●				

	AUS	CAN	FRA	GER	GBR	HK	ISR	JAP	NZ	POL	SPA	USA†
Pop concerts	•					•						
Singing		•										
Theatre	•	•	•		•	•		•		•		
Theatre/concert				•				•				
Making videos								•				
Writing								•				
Home-based activities												
Car repairs						•		•				
Cards, board games		•	•									
Chess, checkers		•	•									
Collecting (stamps, coins)		•										
Computer/video games		•							•			
Crafts/arts		•		•	•	•		•	•			
Do-it-yourself		•	•	•	•			•		•		
Entertaining at home	•	•		•	•			•	•			
Gardening for pleasure	•							•				
Gourmet cooking								•				
Household skills						•						
Indoor games	•											
Listening to music	•	•	•		•	•	•	•	•		•	
Listening to radio					•	•				•	•	
Listening to records/tapes												
Model-making									•			
Outdoor games												•
Playing with pets	•			•								
Playing with children				•								
Reading magazines		•	•	•						•	•	

Table 14.2. *continued*

	AUS	CAN	FRA	GER	GBR	HK	ISR	JAP	NZ	POL	SPA	USA†
Reading books		●	●	●	●	●					●	
Reading newspapers		●	●	●						●		
Reading newspapers/magazines						●				●		
Reading	●					●	●					
Relax, do nothing	●	●	●	●	●	●		●	●			
Sewing, knitting, etc.				●					●			
Spend time at home/with family				●				●			●	
Study								●				
Talk on telephone (15 mins +)												
Watch TV	●	●	●	●	●	●	●		●	●	●	
Woodwork, carpentry	●	●										
Outdoor recreation												
Backpacking								●				●
Beach												
Bird watching	●											
Camping		●				●		●				●
Kite-flying		●				●						●
National park												
Parks								●				
Picnic/barbecue	●					●						
Picnic/hike/nature walk	●					●		●				
Walk in country/bushwalking/hiking	●											
Social/informal recreation				●								
Amusement parks etc.										●		
Auctions		●										

	AUS	CAN	FRA	GER	GBR	HK	ISR	JAP	NZ	POL	SPA	USA†
Bingo	●	●										
Church/religious activities	●	●	●	●		●						
Club/assoc. member				●		●						
Club visit (licensed/night)	●	●				●		●				
Community/voluntary work	●			●		●						
Conversation	●						●					
Dancing, discotheque			●			●	●	●				
Dining out	●	●	●	●		●		●	●			●
Driving for pleasure	●		●			●		●	●			
Electronic/computer games	●	●				●		●				
Excursions	●											
Exhibitions							●	●		●		
Fair or festival		●	●					●				
Gambling		●						●				
General interest courses	●											
Going out/evening			●									
Hobbies	●	●				●	●	●				
Horse races/trots/dog races	●					●		●				
Karaoke			●					●				
Motor sport					●							
Nightclub, disco	●											
Outings		●		●								
Pachinko								●			●	
Pub/cafe/tea house visit	●	●	●	●		●		●	●			
Sauna/massage						●	●					
Shopping for pleasure	●					●			●			
Social activities						●						

Table 14.2. *continued*

	AUS	CAN	FRA	GER	GBR	HK	ISR	JAP	NZ	POL	SPA	USA†
Special interest courses	●					●						●
Sport spectator	●	●	●	●				●		●	●	
Tea ceremony				●				●				
Travelling overseas		●						●				
Visit/be with friends/relatives	●		●		●			●	●		●	●
Visitor centre												●
Walking dog	●	●										
Walking	●	●		●		●			●			●
Working for a church group				●	●	●	●		●			
Sport/physical recreation												
Aerobics	●		●		●	●		●				
Archery/shooting					●	●						
Athletics			●		●	●						
Australian Rules Football												
Badminton					●	●		●				
Ball games						●						
Baseball/softball	●		●			●		●				●
Basketball	●	●	●		●	●		●				●
Bowls/bowling					●	●		●				
Boxing/wrestling					●	●						
Climbing	●	●	●		●	●		●				
Cricket					●	●						
Cricket: outdoor	●											
Cricket: indoor	●											
Curling		●			●							

	AUS	CAN	FRA	GER	GBR	HK	ISR	JAP	NZ	POL	SPA	USAt
Cycling	•		•		•	•		•				•
Darts					•							
Diving	•					•		•				
Exercise, keep fit												
Fencing						•				•		
Fishing/hunting					•			•				
Fishing	•	•	•		•	•		•				•
Football (see also soccer etc)			•		•							
Gateball/croquet	•				•			•				•
Golf	•	•	•		•	•		•				•
Gymnastics		•			•							
Handball					•	•						
Hiking/backpacking												
Hockey		•										
Hockey/lacrosse: outdoor	•		•		•	•		•				•
Hockey/lacrosse: indoor	•					•		•				•
Horse-riding					•							
Ice-skating		•	•		•							
Informal sport									•			•
Jet skiing	•	•	•	•		•		•				•
Jogging/running	•		•					•				
Judo			•					•				
Lawn bowls	•					•						
Martial arts	•		•		•	•		•				
Netball					•	•						
Netball: outdoors	•					•						

Table 14.2. *continued*

	AUS	CAN	FRA	GER	GBR	HK	ISR	JAP	NZ	POL	SPA	USA†
Netball: indoors	•											
Orienteering	•											•
Pinball, pool, shuffleboard		•						•				
Playground games					•	•						
Pool/snooker/billiards	•				•	•						
Rink sports												
Roller-skating			•		•							
Rugby	•		•									
Rugby Union	•	•			•	•						
Rugby League	•											
Sailing		•	•		•	•						
Sailing/canoeing/boating	•				•	•						
Self-defence					•	•						
Shooting/hunting	•	•	•		•	•						
Skateboarding	•				•	•						
Skating		•	•		•	•						
Skating, skiing												
Skiing	•		•			•		•				•
Soccer (see also football)	•					•						•
Soccer: outdoor								•				•
Soccer: indoor								•				•
Softball												
Sport (at least one)			•	•	•	•	•	•	•		•	•
Squash	•				•	•						•
Strength sports (weights, box)		•										

	AUS	CAN	FRA	GER	GBR	HK	ISR	JAP	NZ	POL	SPA	USA†
Surfing/lifesaving	•											•
Swim in own/friends' pool	•											
Swimming	•	•	•			•		•				•
Swimming: outdoor					•							
Swimming: indoor					•							
Swimming: in pool				•								
Swimming: non-pool												•
Table tennis	•		•		•	•		•				•
Team sports		•			•							
Ten-pin bowling		•							•			
Tennis, racquet sports	•		•		•	•		•				•
Tennis	• •		•		•	• •		•				• •
Touch football	•											
Volleyball					•	•						
Water-skiing						•						
Water sport: excl sailing	•		•		•	•		•				
Water activities: non-power			•		•	•		•				
Weightlifting/bodybuilding			•		•							
Windsurfing			•			•						
Yoga					•	•		•				

* Other activities may have been included in surveys but not reported here.

† Data from Table 3.20 included.

such as the arts and sport, which they promote. It is clear that, to date, the surveys have not been fully effective in providing data for assessing the effectiveness of such policies. Much more research, with better data, is required to address the question of trends in participation.

INEQUALITIES

While absolute levels of participation cannot be compared cross-nationally, some of the patterns of relationships between participation and key socio-economic and demographic variables can be compared in an informal way. In most of the contributed chapters the relationships between participation and a number of traditional social variables are examined. Gender, age and socioeconomic status are complex phenomena and their relationships with leisure are the subject of numerous research approaches. The survey data presented here represent just one such contribution to the mosaic of data, theory and interpretation available.

In relation to gender the picture presented is very mixed. Where they are available, time-budget data indicate that women generally have less leisure time than men, particularly women in the paid work force. As regards participation patterns, women tend to be less active in sport and more active in arts and cultural activities, the differences being more marked in some countries than in others. The finding that women also tend to be more active in home-based activities could be interpreted as a result of choice but can also be seen as a reflection of *lack* of freedom of choice, since women's leisure choices are often confined to the home because of child care and domestic responsibilities as well as other economic and cultural constraints.

The surveys generally indicate that the range of leisure activities engaged in declines with age. For a few activities, such as television watching, some arts activities and specific sports such as golf and bowls, participation rates are higher among the older age-groups than among the young – but in general the reverse is the case. Again, time-budget data present a different picture, with retired people, inevitably, having comparatively large amounts of leisure time available. It is widely accepted that an active leisure life can enhance health and the quality of life generally for older people: the picture of *declining* levels of participation with age may therefore be viewed with concern. It could of course be the case that older people choose to engage in fewer activities, but more intensively. But an alternative view is that people who drop leisure activities in middle age, due to family and work commitments, seem not to adopt new activities when leisure time increases upon retirement: the result is an increase in time spent on passive 'time fillers', such as watching television, rather than an intense engagement in an activity of choice (Carpenter, 1992; Dodd, 1994).

As regards socioeconomic status, based on occupation and education, the general picture emerging from the surveys is that there are marked differences in patterns of leisure participation across the socioeconomic spectrum. The economically and educationally advantaged groups in the community generally have higher levels of participation in all activity groups, even though they tend not to have more leisure time. The exception is France, where manual workers have higher participation rates in sport. As with the aged, it is possible that the less advantaged groups adopt a more intensive involvement with fewer activities, but it is also possible that a reduced leisure 'repertoire' results in a lesser quality of life. Again the public policy dimension is relevant: it seems clear that the groups who benefit most from government programmes and subsidies in the arts, sport and outdoor recreation are the economically privileged groups. After several decades of implementing modern public leisure policies, the universal rights to leisure, as discussed at the beginning of this book, are some way from being realized in practice.

PROBLEMS AND ISSUES

Some of the chapter authors found that a number of leisure participation surveys within their countries had been conducted in parallel – but often in relative isolation from one another. A need was identified to synthesize the results of research projects in order to provide an overview of the research directions and leisure trends and characteristics of each participating country and a feel for possible research gaps. As the authors of these chapters searched for answers to the questions and issues we had posed, they also made contributions to the accumulation of knowledge on leisure behaviour they had found to be problematic. Many have made important contributions by synthesizing the work on leisure surveys within their countries.

The on-going problems for national leisure participation surveys might be summarized as the 'three Cs': continuity, comparability and comprehensiveness. Continuity refers to the need to conduct surveys on a regular and frequent basis – every 2 or 3 years at least. Comparability refers to the need to ensure that surveys *within* countries are comparable, but also to consider ways in which comparability *between* countries might be achieved in future. Finally, comprehensiveness is a reference to the need to cover all aspects of leisure and to include as many activities as possible. However, as Zuzanek suggests (Chapter 3), the long-term accumulation of comparable, standardized data sets is hampered by changes of personnel in key government agencies and the desire of new personnel to make their mark on the process by making changes – which result in loss of comparability.

An issue for promoters of national surveys is to consider the respective roles of time-budget and questionnaire survey methodology. Zuzanek points out that time-budget studies are good at dealing with everyday activities, such as television watching and other home-based or frequently engaged-in activities, but are less effective in gathering information on activities which, while they may be key indicators of lifestyle, may not take up a great deal of time on a day-to-day basis – for example theatre going or visiting art galleries. The researcher is in fact faced with a spectrum of approaches related to the 'reference period' used, ranging from 1 or 2 days – effectively a time budget – via a week or a month to twelve months. Beyond 12 months lies the 'personal leisure history' or biographical method, which also has a role to play in leisure research (Hedges, 1986; Smith, 1994). Finding the right balance between these approaches and the advantages and disadvantages they offer, is a challenge for survey researchers around the world.

THE DEVELOPMENT OF COMPARATIVE RESEARCH

The task of the current project was to review the *status quo* and to bring together information on the availability of data and insights into leisure behaviour and leisure trends from a number of countries. It was not our intention for the book to be comparative in the sense of collecting and analysing comparable data sets. However, the book can be seen as representing the early stages of a programme leading to a more comparative international study of leisure.

There is no reason to believe that there exists an easy and straightforward entry into cross-national comparative leisure research. Leisure phenomena are complex and are particularly so in the context of comparative social science. All the eternal and unsolved problems inherent in social science research are unfolded when engaging in cross-national leisure studies. Few of the methodological and theoretical difficulties we have learned to live with can be ignored when we examine critically such questions as what comparative leisure research is, how we go about comparative work, and how we interpret similarities and differences in countries compared. When the focus of the research moves to another analytical level (from single-nation to cross-national) then the problems are more likely to be exacerbated. Yet, interest in cross-national study of leisure remains high, and the need as well as demand for comparisons of leisure patterns across countries is growing.

This is a reflection of the growing internationalization of leisure forms and the concomitant export and import of their social, cultural and economic manifestations across national borders. As discussed in Chapter 1, cultural, entertainment and leisure forms flow between countries in ways we have never seen before. The trend is reflected in the establishment of international leisure and tourism organizations, such as the World Leisure and Recreation Association, Research Committee 13 (Leisure) of the International Sociological Association, and the World Tourism Organization. Cross-national educational programmes are being developed, including the World Leisure and Recreation Association International Centre of Excellence (WICE) and the ERASMUS programme in Europe, and many *ad hoc* arrangements between individual institutions.

Cross-national leisure research may have to shift its emphasis from seeking uniformity among variety in patterns of leisure participation to studying the preservation of enclaves of uniqueness among growing homogeneity and uniformity in leisure.

In the light of these perspectives on cross-national leisure research, three central types of issues, namely theoretical, methodological and organizational issues, are crystallized and analysed briefly in the remainder of this chapter.

THEORETICAL AND METHODOLOGICAL ISSUES

The belief that the complexities of social phenomena increase in comparative social science is part of collective social science folklore. There is, nevertheless, a sense that, even among the complexities and uncertainties of contemporary society, some predictabilities exist in leisure phenomena and that there are many similarities in leisure participation patterns among countries in the economically developed world. This sense of order-in-complexity is very strong in comparative social science, but the challenge is to summarize, in a theoretically or substantively meaningful way, the order that seems apparent across a diverse range of countries.

One of the main criticisms levelled at cross-national leisure research is that there is too heavy a reliance on survey methodology and quantitative analysis and this has frequently dictated the problems and questions studied and resulted in what Iso-Ahola (1986, p. 368) has called 'the undirected accumulation of empirical data'. It is argued that 'number crunching', in the form of statistics, is often a substitute for establishing a meaningful dialogue between ideas and evidence and the quest for comparison of mainly quantitative data may have been at the expense of theoretical rigour. It is claimed that the outcome has been:

> . . . too many research reports which confront the reader with a mass of
> unstructured facts. The attempt to provide over-comprehensive results ends in
> interesting information being drowned in a welter of data of lesser interest.
> (Coenen-Huther, quoted in Aitchison, 1993)

It is not suggested that these criticisms have been fully addressed in this volume; indeed, some of the implied criticisms of the survey method, for example that the sheer quantity of data makes it difficult to 'see the wood for the trees', can probably never be fully overcome. However, the *potential* exists for addressing these issues and for delineating an acceptable role for national participation surveys. First, it should be stressed that large-scale national leisure surveys tend to be conducted for policy rather than theoretical purposes (Veal and Cushman, 1993); while policy-research issues are of interest in themselves, in fact much of the use of survey data in leisure studies is opportunistic, secondary analysis. This volume is an example of such an opportunistic undertaking. Participation surveys can play a role in developing theory, but they are rarely designed for that purpose. Second, surveys do have *some* role to play in examination of focal issues of concern to social researchers. Among these are inequality in relation to gender, age and socio-economic status, as discussed above, and in relation to other factors such as ethnicity and race, which are not specifically addressed in this volume (except in the US chapter). That leisure survey research can engage with issues of social change is illustrated particularly by the chapters on Poland and Japan, two societies experiencing very different sorts of social, political and economic change. Indeed, as we have attempted to illustrate in this and the introductory chapter, one of the driving forces behind the search for reliable time-series data on leisure participation is an attempt to address the question of whether leisure is one of the 'dividends' of economic development, as conventional wisdom would have it.

Real difficulties nevertheless remain in undertaking such a task on a cross-national basis. One of the most fundamental theoretical and methodological issues in cross-national leisure is whether the concepts employed in the analysis are truly equivalent. To obviate the possibility that differences in findings are merely an artefact of differences in concepts and methods – in the nature of the samples, in the meaning of the questions asked, in the completeness of the data, in measurement – comparative researchers try to design the studies to be comparable, to establish both linguistic and conceptual equivalence in questions and in coding answers, and to establish truly equivalent indices of the underlying concepts (Scheuch, 1968). Edward Suchman (1964, p. 135) stated the challenge as follows: 'A good design for the collection of comparative data should permit one to overcome as much as possible that the differences (and similarities) observed ... cannot be attributed to the differences in the (concepts) and methods used.' Unfortunately, one can never be certain that the challenge has been met.

The issue of equivalence of concepts in comparative research is complex (see, for example, Kohn (1989)) and is no less a challenge in cross-national leisure research. At the analytical level, researchers engaged in cross-national comparisons are warned against cultural bias and the dangers of interference from their own cultural values. As illustrated by Hantrais and Samuel (1991), when new data are being collected the research design should ideally be replicated and the same concepts used simultaneously in two or more countries in matched groups, as is possible when new empirical research is conducted (see Greenfield, 1986; Fell, 1990). However, the problem of 'equivalence' becomes ' ... more difficult to resolve when collaborators from different cultural and linguistic backgrounds are working together across national boundaries and when agreement has to be reached over parameters and the functional equivalence of concepts . . . ' (Hantrais and Samuel, 1991, p. 393).

Roberts' (1994) contention that leisure researchers should not attempt to integrate substantially different conceptualizations of leisure (time and experience) may be a useful step towards suggesting solutions to some of the theoretical issues associated with the study of leisure in intersocietal comparisons. He has demonstrated why researchers should not adopt conceptual frameworks which define leisure both as a type of time and as a type of experience and the reasons this conceptual confusion has tied the sociology of leisure into conceptual knots that prevent the discipline engaging its subject matter. Particularly relevant for cross-national leisure research is Roberts' argument that this conceptual confusion in studying leisure phenomena has been a factor in handicapping the sociology of leisure engaging in, and helping to explain, various major social and economic trends, such as globalization, the transition to post-Fordism, the emergent post-modern condition and socioeconomic polarisation in western societies.

Another challenge facing cross-national survey research is how to locate the leisure phenomena under study within their wider social context. This requires the incorporation of often qualitative material along with the largely quantitative material produced by the survey process. According to Hantrais and Samuel (1991), this approach has become widely accepted amongst comparative social scientists, and it is outlined by Bramham *et al.* as follows:

> Instead of thinking of comparative research as an analysis of particular isolated cases, it would perhaps be better to speak more generally of studies of systems interaction, which deal with the way in which phenomena (groups, collectivities, practices) are related to one another and where these are situated in historical time and global space. Such an approach would draw our attention to relations of presence and absence, to the ways in which certain styles of urban leisure policy are related not just to local circumstances, but perhaps through these, to more national, global or historical processes.
>
> (Bramham *et al.*, 1989, p. 286).

In a particular national setting, the prevailing economic circumstances and political conditions are always of overwhelming importance in determining the manner in which the discourse surrounding leisure is conducted (see Henry, 1988; Mercer, 1994). Viewed in this way one would, for example, anticipate quite different discourses in fully industrialized as opposed to partially industrialized societies, arising from different histories, and differing social, economic and political environments. Marie-Therese Letablier, however, argues that this opens up a whole new dimension for research:

> By asking questions about comparability we are raising others . . . what is being compared is not so much the phenomena which are initially to be studied but

the contexts in which they are located. Rather than being a difficulty which has
to be overcome, comparability therefore becomes an object of analysis in its
own right. This implies that the reader needs to go back in time and take
account of the historical dimension of the phenomena under investigation.
                                          (Letablier – quoted in Hantrais and Samuel, 1991)

Walter Tokarski's (1994) proposed project, *Leisure Structures in Europe* is
a good example of comparative research in which leisure is to be analysed in
its wider historical and social context and with respect to cross-national
differences and similarities. Qualitative and quantitative methods will be
utilized to examine, in comparative perspective, the structural elements of
leisure as a social system throughout Europe; the actual and potential role of
leisure as an economic factor in different European regions; and the role of
leisure in shaping sociocultural identity.

THE FUTURE

Comparative analysis in leisure is attracting growing interest, as reflected in
the emergence of cross-national educational programmes and the growing
number of research publications. The globalization of leisure issues and
problems has been placed on the international agenda through such media as
the Brundtland Report on *Our Common Future* (1987) and the United Nations
World Summits on the Environment (1992) and on Social Development
(1994). A national leisure issue is seldom merely national any more. The air
we breathe is polluted from faraway sources. An increase or decline in the
value of the United States dollar in relation to the currencies of other countries
has immediate implications for leisure expenditure, consumption and participation worldwide. Within such a perspective the use of the nation-state as the
unit in leisure research is limiting – hence the appeal of comparative studies of
leisure.

Notwithstanding the important role of international organizations in
facilitating cross-national research, the barriers to the organization of comparative research are formidable. These include the theoretical and methodological considerations mentioned above, but also practical problems associated with obtaining sufficient resources, in terms of money, time and
personnel, to conduct the research. It is difficult to suggest solutions to these
pragmatic and logistical issues which create barriers to organizing cross-
national leisure research. So far most of the cross-national studies on leisure
have been located in Western Europe and North America. This is also where
most of the leisure researchers, institutions, data banks, and agencies for
funding leisure research are located. The research in the 1960s and most of
the 1970s of the International Social Science Council (ISSC) and the European
Centre for Research and Documentation in the Social Sciences, especially

through the work of Stein Rokan and Alexander Szalai, did much to promote comparative social research, including comparative leisure research (see Szalai, 1972; Scheuch, 1990). This earlier work was concerned with addressing issues about the apparent trajectories of societies towards the development of a grand theory and universal truths in leisure sociology. More recent work arising out of the Research Committees of the International Sociological Association has tended to focus on particular issues and problem areas through broadening the base observations. There may be considerable benefit in re-visiting the methods adopted by these earlier comparative research organizations and, in particular, how they overcame barriers to the conduct of cross-national research. Among the activities of the Standing Committees for Comparative Research (1967–78) of the ISSC was the organization of international conferences on comparative research, the development of a network of researchers, the facilitation of data centres and archives, the *post-hoc* examination of a number of large-scale comparative studies and, perhaps most importantly, the organization and conduct of seven international workshops and seminars and three European Summer Schools on comparative social research.

This volume has drawn on participation surveys as the most convenient entrée into cross-national research. It has been based on existing data, acknowledging all the limitations that this inevitably places on any attempts at comparison. In future developments in cross-national leisure research, it is to be hoped that, in addition to an extension of this methodology, in terms of increased comparability and wider geographical coverage, we will also see the incorporation of other methodologies, including historical, qualitative and experimental methods.

While acknowledging these problems and barriers, the database for analysing leisure participation in economically developed nations has been improved substantially in the last two decades. The growth in the number of leisure scholars with international and cross-national interests and viewpoints has contributed significantly to an international division of labour. Much has been learned and is being learned about ways and means of collaboration in the conduct of these projects, with inevitable mistakes, and the norms of reciprocity are evolving rapidly. A challenge for the future is to incorporate the knowledge of, and expertise in, leisure participation from scholars of developing countries as this may stimulate new insights into an increasingly global phenomenon.

REFERENCES

Aitchison, C. (1993) Comparing leisure in different worlds: uses and abuses of comparative analysis in leisure and tourism. In: Collins, M. (ed.) *Leisure in*

Industrial and Post-Industrial Societies. Eastbourne, Sussex, Leisure Studies Association, pp. 385–400.

Bramham, P., Henry, I., Mommas, H. and Van Der Poel, H. (eds) (1989) *Leisure and Urban Processes: Critical Studies of Leisure Policy in Western European Cities.* Routledge, London.

Bruntland (1987) *Our Common Future.* Oxford University Press, Oxford.

Carpenter, G. (1992) Adult perceptions of leisure: life experiences and life structure. *Society and Leisure* 15(2), 587–606.

Darcy, S. (1993) Leisure participation in Australia: the monthly data. *Australian Journal of Leisure and Recreation* 4(1), 26–32.

Dodd, J. (1994) The male leisure repertoire, before and after the transition to mid-life. In: Simpson, C. and Gidlow, B. (eds) *Leisure Connections.* Proceedings of the Australian and New Zealand Association for Leisure Studies Second Conference, Canterbury. Lincoln University, New Zealand, pp. 51–56.

Fell, J. (1990) The Influence of Leisure on Work in Britain, West Germany, France and Japan: A Cross Cultural Comparative Study. PhD thesis, Aston University, Birmingham.

Greenfield, S.C. (1986) Middle-Class Families and their Leisure Patterns in Britain and France. PhD Thesis, at Aston University, Birmingham.

Hantrais, L. and Kamphorst, T.J. (eds) (1987) *Trends in the Arts: A Multinational Perspective.* Giordano Bruno Amersfoot, Voorthuizen, Netherlands.

Hantrais, L. and Samuel, N. (1991) The state of the art in comparative studies in leisure. *Loisir et Société/Society and Leisure* 14(2), 381–398.

Hedges, B. (1986) *Personal Leisure Histories.* Sports Council/ESRC, London.

Henry, I. (1988) Alternative futures for the public leisure service. In: Bennington, J. and White, J. (eds) *The Future of Leisure Services.* Longman, Harlow, Essex, pp. 207–244.

Iso-Ahola, S. (1986) Theory of substitutability of leisure behaviour. *Leisure Sciences* 8(4), pp. 367–389.

Kohn, M. (1989) Cross-national research as an analytical strategy. In: Kohn, M. (ed.) *Cross-National Research in Sociology.* Sage, Newbury Park, CA, pp. 77–102.

Mercer, D. (1994) Monitoring the spectator society: an overview of research and policy issues. In: Mercer, D. (ed.) *New Viewpoints in Australian Outdoor Recreation Research and Planning.* Hepper/Marriott Publications, Melbourne, pp. 1–28.

Roberts, K. (1994) Untying leisure's conceptual knot: returning to a residual time definition. In: *Research Committee 13, International Sociological Association Newsletter,* November, pp. 3–10.

Scheuch, E (1968) The cross-cultural use of sample surveys: problems of comparability. In: Rokan, S. (ed.) *Comparative Research Across Cultures and Nations.* Mouton, Paris, pp. 19–37.

Scheuch, E. (1990) The development of comparative research: towards a causal explanation. In: Oyen, E. (ed.) *Comparative Methodology: Theory and Practice in International Social Research.* Sage, Newbury Park, CA, pp. 179–206.

Smith, L.M. (1994) Biographical method. In: Denzin, N.K. and Lincoln. Y.S. (eds) *Handbook of Qualitative Research,* Sage, Thousand Oaks, CA, pp. 286–305.

Suchman, E. (1964) The comparative method in social research. *Rural Sociology* 29, 123–137.

Szalai, A. *et al.* (1972) *The Use of Time.* Mouton, The Hague.

Tokarski, W. (1994) Leisure systems in Europe and their economic and cultural impact in comparative perspective. In: Henry, I. (ed.) *Leisure in Different Worlds*, Volume 1. Leisure Studies Association, Eastbourne, UK, pp. 173–178.

Veal, A.J. and Cushman, G. (1993) Leisure in different worlds: the survey evidence. In: Collins, M. (ed.) *Leisure in Industrial and Post-Industrial Societies*, Eastbourne, Sussex, Leisure Studies Association, pp. 401–412.

Index